TABLE OF CONTENTS

1. PLANNING GUIDELINES

This book can help you make good decisions while planning farm grain handling, drying, and storage systems. Planning involves **selecting** specific equipment and their capacities, **organizing** them into a system, and properly **locating** the system.

Read the entire book to understand the whys and hows of developing a total grain storage system. Then re-read, study, and apply the sections that interest you most. You may also want to consult experienced engineers, knowledgeable equipment suppliers, and producers with systems similar to your plan.

The importance of planning cannot be overemphasized. It requires considerable time, effort, and thought. Good planning results in a system that works well now and can expand for your future needs. A well planned system usually costs no more than a poorly planned one.

This book is for farmers, equipment dealers, and others interested in developing grain systems. It applies to all dry grains. Recommendations are based on research, when available. There is more information on corn than on other grains, because there has been more research. Where research is lacking, guidelines are from the experience and judgement of agricultural engineers representing 12 North Central states.

The following subjects are **not** covered in detail in this book, although they are important:

- **Operation and management.** This book includes mainly what is needed for planning. For operating and management techniques not discussed, consult owner's manuals, county extension agents, and state extension agricultural engineers.
- **Harvesting equipment** capacity is discussed only for its effect on grain receiving.
- **Transportation** from field to storage and from storage to market or feedlot is mentioned only for locating the system or sizing equipment.
- **Feed processing** is not discussed, although you need to consider how grinding, weighing, and mixing fit into the grain system.
- **High moisture grain** is excellent for storing livestock feed. Consider handling and storage equipment needed to include it in the dry grain system.
- **Evaluate economics** before construction begins. Your goal is the most functional system that is also economical.

Safety

Another goal is **safe facilities.** Common hazards include pulleys, chains, conveyors, falls, large vehicles, dust, and noise. Unsafe facilities are not acceptable no matter how functional or economical they are. Develop your grain system with the goal of protecting everyone—operator, employees, visitors, family members, and especially children. Do not tolerate safety hazards. Post warnings near potential hazards.

Automatic equipment that can start unexpectedly is especially hazardous. Post signs on automatic equipment, warning that it may start at any time.

Install protective shields on operating equipment. Install lock-out devices on electrical service to bin unloading equipment. Place warning signs on operating switches and bin entrances whenever equipment is being serviced. Talk with your family, coworkers, and friends about safe grain handling practices, especially when seasonal activity begins. Emphasize awareness and knowledge of hazards and risky operating procedures. Wear a dust mask in a dusty facility, particularly when working in a bin. Breathing mold spores from deteriorating grain can be a serious health hazard. Provide masks and install signs at strategic locations reminding workers to use them.

Safety considerations are important when planning any facility. Other safety ideas are addressed throughout the book.

Long-Range Planning

Grain systems include grain handling, holding, drying, cooling, and storage equipment and facilities that must operate as a coordinated unit. The goal is to receive, dry, and transfer grain to storage or loadout, while maintaining quality without impeding grain harvesting or shipment.

You may need to change your grain system to include: increased storage, a bucket elevator, pneumatic conveying, wet grain receiving, a new drying method, a switch to livestock or cash grain, or feed processing. Incorporate present needs into a long-range plan with provisions for growth and improvement.

First evaluate present needs; then add possible growth and change for 3 to 5 years into the future. Then **double the volume of grain handled** and **replan** for this expanded need. The idea is not ridiculous; not everyone will double their production, but doubling the plan forces you to evaluate what might change within the probable facility life.

Most well designed grain and livestock facilities can be doubled and still perform well. If you plan for 3 to 5 years in the future, a doubled unit may serve 6 to 10 years. If the doubled plan is good, it likely can be doubled again without serious compromise, and be useful up to 20 years—a reasonable planning base for most building facilities.

Farm structures have a long life—at least 20 years. The productive life of a farm manager is perhaps 40 years, so you may have only two chances (every 20 to 25 years) to correct a planning mistake.

Collect company literature and equipment specifications. The equipment must handle present needs and fit future plans. While deciding on the type of system you need, visit farms with different systems and equipment. Look at alternatives, advantages, and disadvantages. After selecting a system type, visit systems of comparable size with similar equipment. Ask what works best; what causes problems; what changes the operator would make; and what needs the most repairs. In addition to the operator, talk to others who may be candid: a partner, son, daughter, or employee.

Select the general types of equipment and sketch the system. Sketches are easy to change, but a concrete foundation is hard to move. See Chapter 5 for more about site selection, arrangement, sketching, and system evaluation.

Consider the effect of changes in grain or livestock volume on timeliness, capacity to get the job done, grain quality, and dependability of services (e.g. custom harvesting and hired help). Plan for expected volumes of each grain produced. Analyze the system for grain production changes, feeding vs. marketing, etc.

Though larger storage bins generally cost less per bushel, smaller bins add flexibility for different grains or grain qualities. Cropping patterns may change or some grain may be moldy at harvest, so consider at least two bins for each type of grain.

With a young family and limited resources, it makes sense to keep it simple and rely on more manual work. But, when you are 10 years older and 20 years more tired, you may want additional mechanization. Ask yourself if more mechanization is wise now, and if it can be easily added later.

Planning principles for large and small systems are the same. The main differences are grain flowrates, drying capacity, storage capacity, and automation level.

Grain Quality

Grain quality at harvest is influenced by grain variety, weather, and harvester adjustment. Minimizing grain damage to maintain grain quality requires good drying, cooling, and handling equipment, and conscientious stored grain management. High quality grain:

- Has few fines and foreign materials.
- Has little physical damage and few stress cracks.
- Has little mold and insect damage.

Clean grain dries and stores better. Install a grain cleaner to remove fines and foreign material or at least plan space for one. Allow for bypassing the cleaner when cleaning is not necessary. Make BCFM (broken corn and foreign material) or dockage (small grains) tests on incoming grain to help evaluate harvesting and variety performance. You can buy or build suitable grain grading screens.

Drying Damage

High temperature grain drying (especially corn) followed by rapid cooling produces stress cracks. Stress cracks are fractures inside the kernel that may not come through the outer seed coat. Cracking is greatest for corn when it is rapidly dried through the critical 19%-14% moisture range and then is quickly cooled. Stress cracks reduce milling quality and increase kernel breakage during handling. Slow drying and delayed cooling reduce stress cracking.

Other potential problems caused by high drying temperatures are:

- Reduced baking and milling quality of wheat.
- Reduced germination of malting barley.
- Discoloration (browning) of kernels (heat damage).
- Fractured soybeans, when drying air relative humidity is below 40%.

Desired grain quality limits drying air temperature. Only broad guidelines can be given for recommended drying air temperatures, because many factors are involved.

Physical grain quality is actually determined by final **grain** temperature, which depends on drying air temperature, airflow rate, initial and final grain moisture content, and drying air distribution through the grain. Optimum drying air temperatures depend on dryer type, airflow rate, grain type, end use (feed, market, seed), and initial and desired final grain moisture content and quality.

For any high temperature grain dryer, lower drying air temperatures maintain higher physical grain quality. Physical grain quality includes test weight, BCFM, color, and brittleness. Very high initial moisture content or drying to excessively low moisture content requires lower drying air temperatures to maintain grain quality. Poor initial grain quality can also require lower drying air temperatures. Generally, to maximize drying rate and fuel efficiency, dry at as high a temperature as you can and still be satisfied with the final grain quality. If quality is not satisfactory, reduce drying air temperature for any type of heated air dryer.

Drying Temperatures

- For seed grain, dry at temperatures below 110 F regardless of all other factors.
- Dry malting barley in high airflow automatic batch or continuous flow dryers up to 120-130 F.
- Dry edible beans using drying air temperatures below 100 F to reduce splits and seed coat cracking.
- Dry market soybeans at temperatures up to 140 F in high airflow column dryers, but not over 110 F in bin dryers. At these temperatures, splits and seed coat cracking do not generally affect marketability. However, frequently check bean quality during drying, and reduce temperature if necessary.
- Dry small grains (wheat, durum, rye, etc.) for milling below 160 F in high airflow batch and continuous flow dryers, and below 120 F in bin dryers.

Feed grains such as corn and milo can be dried at higher air temperatures.

- High speed automatic batch and continuous flow dryers often operate at 200-220 F for corn if cooled immediately in the dryer, and up to 240-250 F with delayed cooling.
- Continuous flow and roof bin-batch dryers operate at 140-180 F.
- Heated air batch bins, with or without stirring, operate at 110-140 F on feed grains. If drying grain at depths exceeding 4′ without recirculating or stirring, limit drying air temperatures to 10 F above ambient air to prevent overdrying.
- Continuous flow dryers with concurrent grain and airflow (grain and airflow in the same direction) may be at 250-300 F because the hottest air is always in contact with the wettest grain. The air temperature and grain moisture content decrease as the air and grain move through the dryer, so the grain is stressed less.

Handling Damage

Increased grain breakage results from improperly installed or operated handling equipment, grain impact at high velocities, and kernels stress-cracked during drying and/or cooling. Very dry or cold grain is more susceptible to breakage. Minimize damage with proper sizing and operation of handling equipment, low drop heights, and low grain velocities. Design the drying and cooling systems to minimize stress cracking, so grain is less susceptible to breakage during handling.

Operate augers at rated capacity to reduce grain damage, especially tube augers without internal bearings. With variable incoming flowrates, reducing auger rpm can keep augers operating at rated capacity. Choose bearing-supported augers where variable incoming flowrates make it difficult to operate augers at rated capacity at all times. Another option is to add a surge (accumulating) bin over the hopper to feed the auger, keeping it full.

Bucket elevators cause little grain damage unless there are large drop heights. When drop heights exceed 40′, grain damage significantly increases. Install grain decelerators in the down spout every 40′ to slow the grain unless impact damage is of no concern.

Pneumatic conveyors cause little damage unless improperly installed or operated. Avoid excessive grain velocities, rough or nonaligned tube joints, sharp turns, and large drop heights.

Storage Damage

Most damage during storage is caused by mold and insects. Mold and insects grow best in improperly dried, damaged, or improperly managed grain. Grain temperature must be controlled with aeration. Pockets of fines (broken kernels, weed seeds, trash, etc.) can cause spoilage because they restrict airflow and provide food for insects and molds. Clean grain to remove fines and adjust grain spreaders to reduce fines concentrations. See the section on aeration.

2. HANDLING

Well organized handling components make a grain storage system workable. Components include conveyors, hoppers, pits, surge bins, spreaders, sweep augers, and cleaners. They receive grain, move it to and from any part of the grain center, and load out the grain.

The most important factors to consider are: performance, capacity, and convenience. Performance means getting the overall job done—not just having enough capacity to handle daily harvest and marketing, but also the production of quality grain with a practical level of supervision. Match overall grain center capacity to daily and seasonal farm needs. Size conveyors to keep up with harvesting and drying.

Convenient Grain Handling

Electrically powered, permanently installed conveyors are most convenient. They move grain at any time to any place despite weather or other farm activities. Because no setup time is necessary, you can schedule low capacity grain conveying needs as easily as high capacity needs. With automatic control, conveyor capacities can be lower because continual supervision is not required.

Plan to eventually mechanize all grain flow. You can start with portable augers to reduce costs, but plan for permanent conveyors as you get older, original equipment wears out, and grain volume increases. Good planning allows you to grow gradually from low investment to a more ready-to-go, convenient system as profits and growth permit.

Develop the receiving/loadout area as the hub of the grain center. Plan equipment so grain can go from the hub to any component (e.g. wet holding, dryer, storage) in the grain center and from any component back to the hub. Bucket elevators are often the hub for larger grain centers, Fig 2-1.

In a small center, grain may not need to return to the hub from every component. But as the system becomes larger and more complicated, return conveyors become more important. If return conveyors are not included originally, plan the system so they can be easily added later.

Equipment Alternatives

Grain centers include wet holding bins, high speed dryers, drying bins, cooling bins, storage bins, cleaning equipment, and surge bins. Some processes are combined: a batch or continuous flow dryer can dry and cool; a stirring bin can dry, cool, and store. And, one process may be in two units: combination drying starts in a high temperature dryer and finishes in a low temperature bin dryer. A component planned for one purpose may have alternative uses that do not seriously impair the original use: an elevated hopper bottom bin can be used for both wet holding and dry grain loadout.

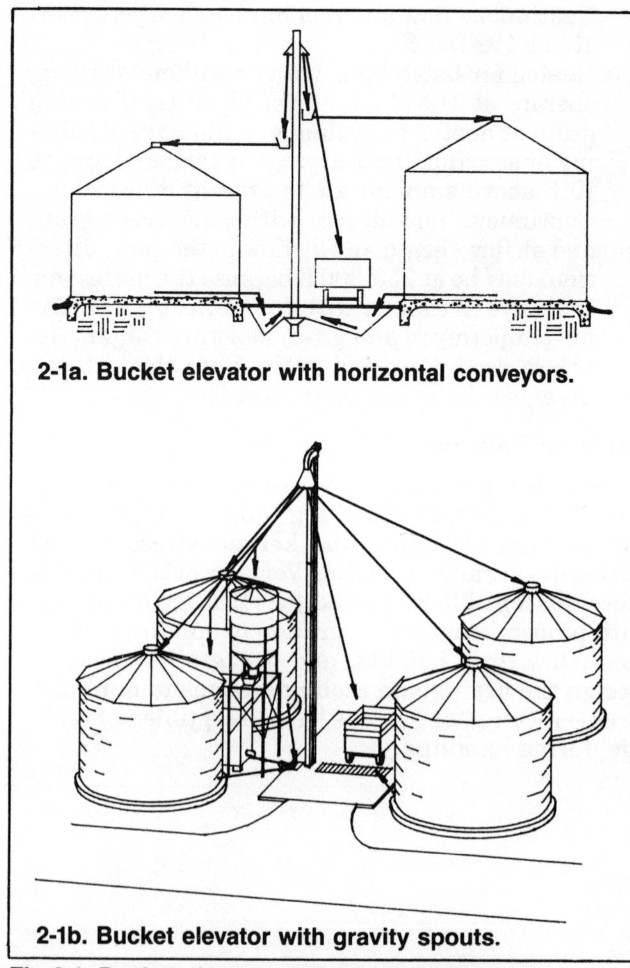

2-1a. Bucket elevator with horizontal conveyors.

2-1b. Bucket elevator with gravity spouts.

Fig 2-1. Bucket elevator at the grain center hub.
Grain from the dump pit goes through the bucket elevator to any grain center component (e.g. wet holding, dryer, storage). Return conveyors move grain back to the hub for loadout or transfer to another component.

Grain flow diagrams or sketches help prevent mistakes during early planning. Fig 2-2 shows common grain center functions and interconnections. The arrows represent conveyors. Table 2-1 is a list of equipment choices for the main functions in Fig 2-2.

Grain Conveying Capacity

Match conveyor capacity to present needs, but allow for increasing needs. Consider buying a larger conveyor and running it full at a lower speed; speed it up later to increase capacity. Compared with a small, high speed unit, a larger, slower conveyor lasts longer, has better power efficiency, and damages grain less.

When one conveyor discharges into another, operate the second at higher capacity (5%-10%) than the first (multiply by 1.05 to 1.10), so the second one does

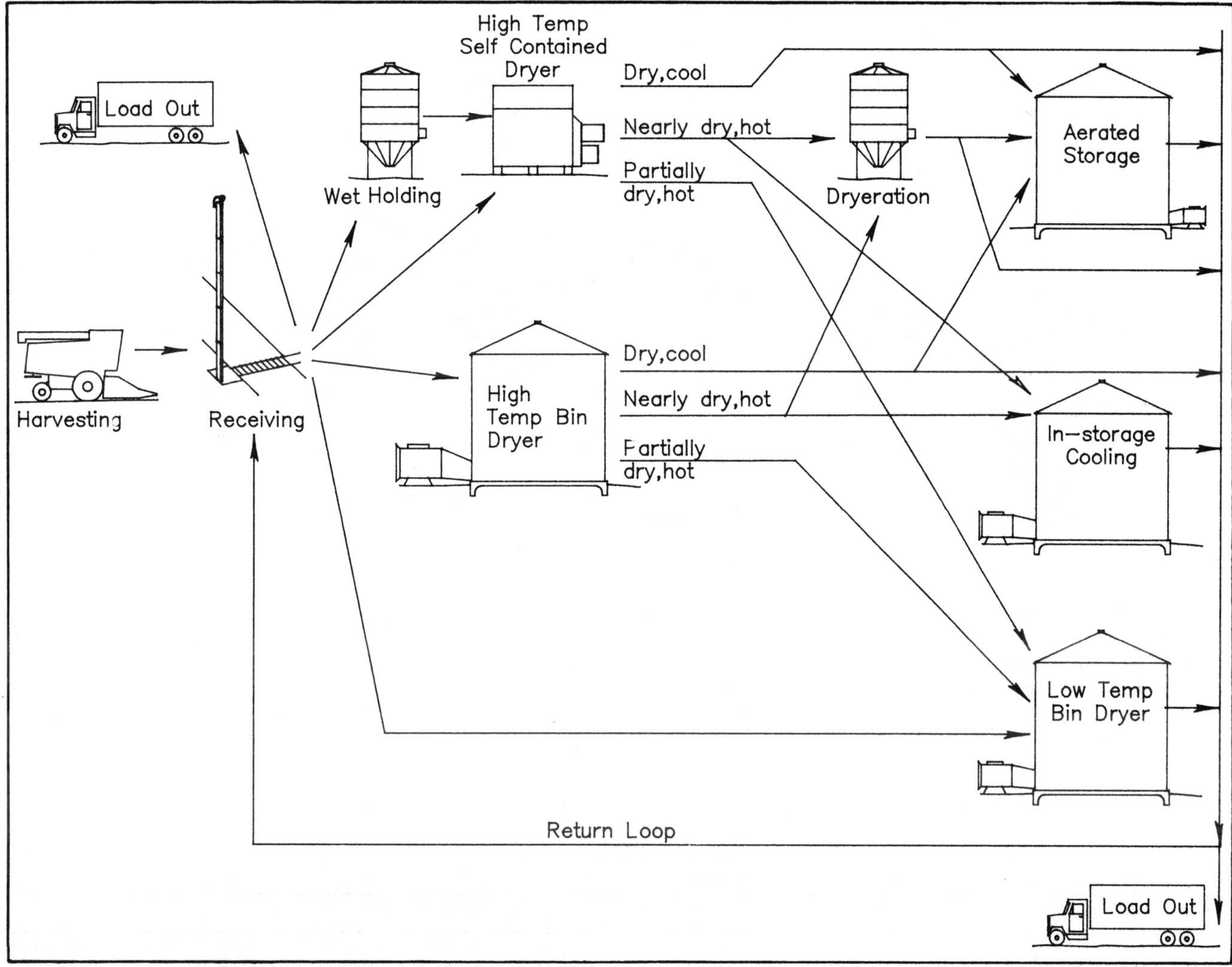

Fig 2-2. Grain center facilities and conveying processes.
Arrows represent conveyors.

not plug. For example, discharge a 1,000 bu/hr conveyor into one with 1,050 to 1,100 bu/hr capacity.

Electrical pressure or flow switches can be used to prevent plugging or overflow when one conveyor discharges into another. Install two switches in the second conveyor's hopper—a high one to shut off the first conveyor when the hopper is full and a low one to turn on the first conveyor when the hopper is almost empty. These switches are less expensive than cleaning spilled grain and unplugging conveyors. Motors overheat if cycled on and off frequently, so make the hopper large enough or the flowrate difference small enough so the motors run at least 4 min per cycle.

You may need both high and low capacity conveying. Receiving and loadout require high capacity conveying, but flow from a continuous flow dryer is relatively low. Widely different rates require either different capacity conveyors or surge bins. Use electrical switches to control flow into and out of surge bins. Surge bin uses include:

- Dump pits, so a truck or wagon unloads quickly and the pit empties over a longer period.
- A batch dryer unloads quickly into a surge bin which delivers grain to storage more slowly.
- Overhead bins fill with low capacity conveyors, but empty quickly by gravity into a dryer or truck.
- A surge bin between a continuous flow dryer and the dry grain conveyor keeps the conveyor (particularly an auger) full to reduce grain damage. It can also accumulate dry grain while the conveyor moves wet grain.

Surge bins are below ground pits, gravity flow wagon boxes, small storage bins, or commercial steel hopper bottom bins. Hopper bottoms cost more but completely unload without mechanical equipment.

Gravity discharge is fast and convenient, but is not free. Gravity unloading from a surge bin requires elevating grain higher, Fig 2-3a, than direct delivery to the same elevation by conveyor, Fig 2-3b. Elevated bins are expensive. Use them where gravity flow has a big advantage and the bin is refilled many times a year. Examples include wet holding bins over dryers, bins over feed grinders, and bins to load trucks.

Commercial hopper bottom bins are rarely justified for bins unloaded only once or twice a year. A bin sweep unloader in a flat bottom bin does the same job without the space loss and high cost of the hopper bottom.

Table 2-1. Equipment alternatives.

Receiving	Wet holding	Drying	Cooling after drying	Storage	Conveying
Conveyor in a separate hopper	Ground-level hopper-bottom tank	Low temperature bin dryer	Cooling in dryer	Round metal bins	Augers
Hopper attached to conveyor	Elevated hopper-bottom tank	High temperature bin-batch dryer	Storages equipped for cooling	Flat storages	Bucket elevators
Portable, self-powered hopper	Flat bottom grain bin with adequate airflow.	High temperature continuous-flow bin dryer	Bins or tanks equipped for dryeration cooling	Converted corn cribs	Flight elevators
Hinged or swinging conveyor attached to primary conveyor	Gravity wagon	High temperature self-contained batch dryer		Converted silos	Pneumatic conveyors
Drive-over conveyor		High temperature self-contained continuous-flow dryer			Bulk or mass flow conveyors
Drive-over pit with conveyor		Combination high temperature/ low temperature dryers			Belt conveyors

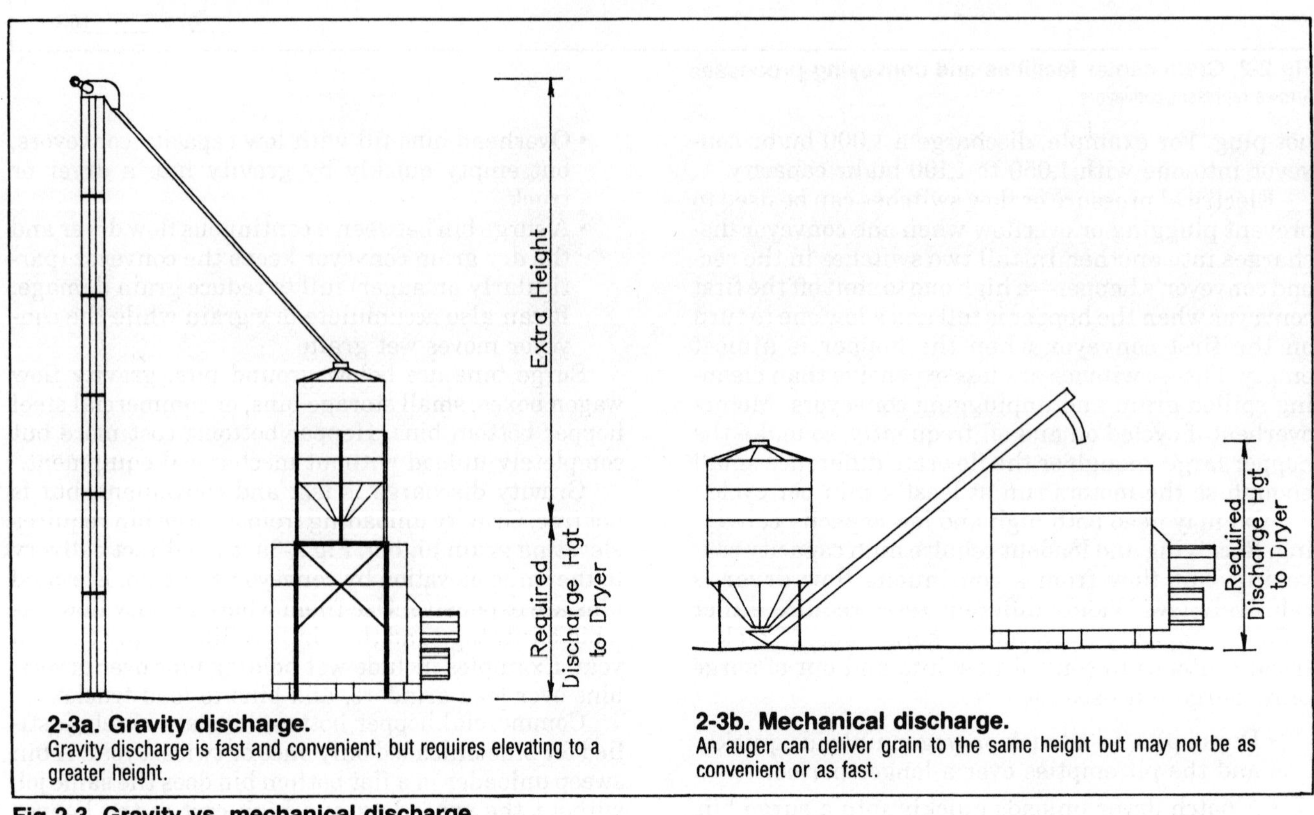

2-3a. Gravity discharge.
Gravity discharge is fast and convenient, but requires elevating to a greater height.

2-3b. Mechanical discharge.
An auger can deliver grain to the same height but may not be as convenient or as fast.

Fig 2-3. Gravity vs. mechanical discharge.

Grain Handling Components

Grain Receiving

Harvested grain enters the grain center in the receiving area and is transferred to a wet holding bin, dryer, or perhaps to storage. Grain receiving is often a bottleneck—provide high receiving capacity to speed truck or wagon unloading.

Above-ground receiving hoppers usually cost less than recessed or drive-over units, Fig 2-4. However, they must be moved for each grain load or backed up to, and spilled grain has to be scooped into the hopper.

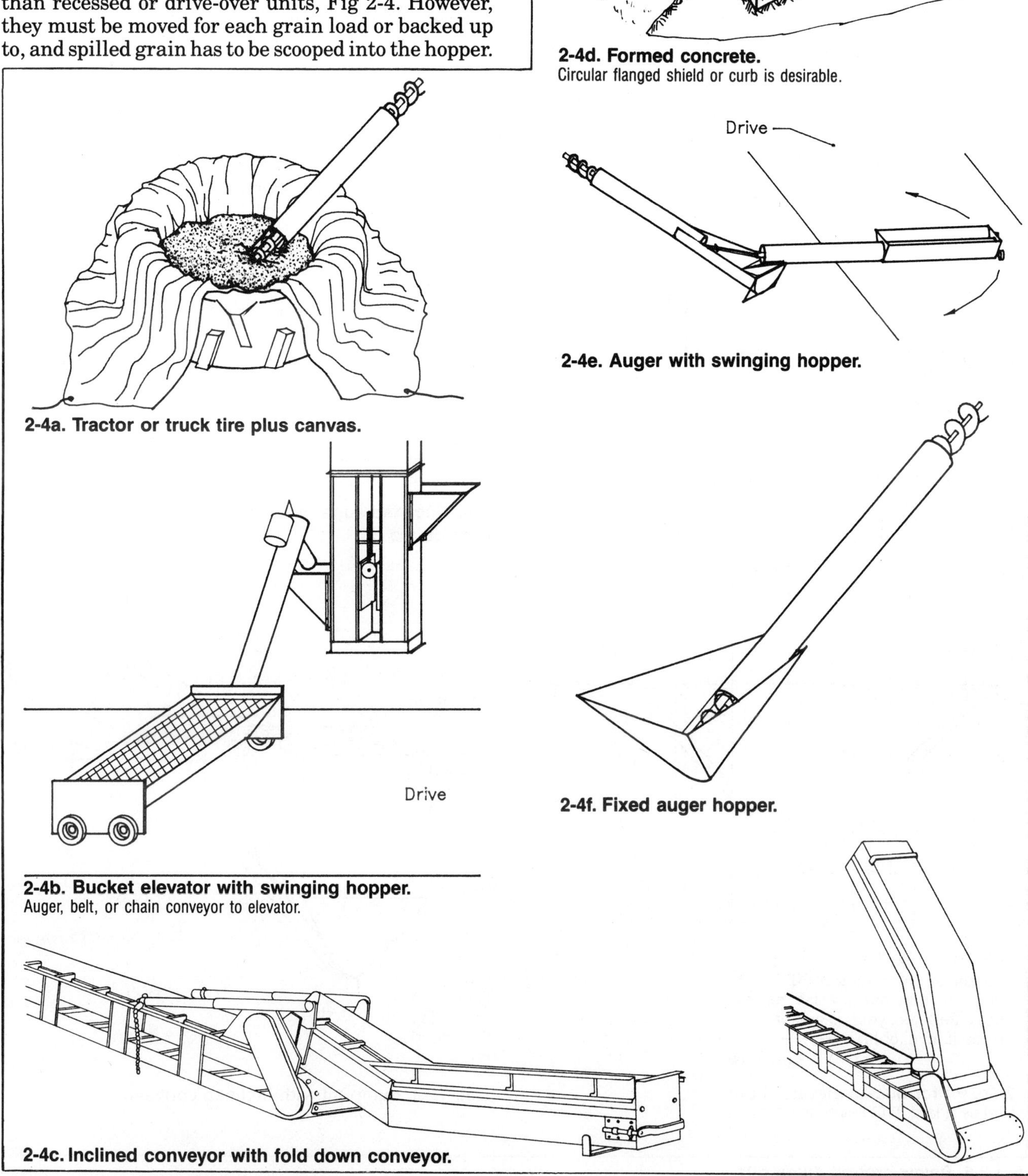

2-4a. Tractor or truck tire plus canvas.

2-4b. Bucket elevator with swinging hopper.
Auger, belt, or chain conveyor to elevator.

2-4c. Inclined conveyor with fold down conveyor.

2-4d. Formed concrete.
Circular flanged shield or curb is desirable.

2-4e. Auger with swinging hopper.

2-4f. Fixed auger hopper.

Fig 2-4. Above-ground dump facilities.

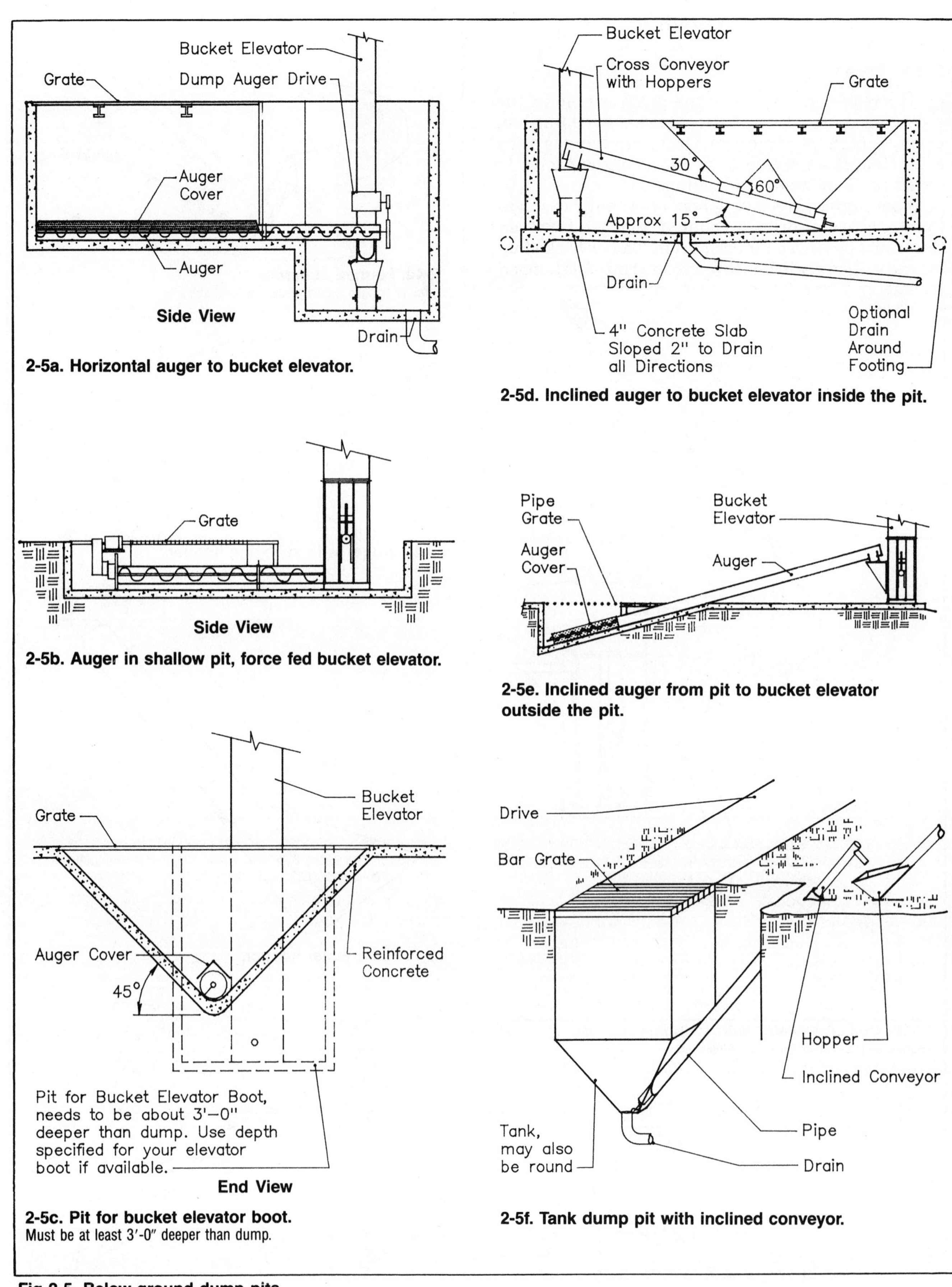

Fig 2-5. Below-ground dump pits.
Dump pits can be concrete or welded steel. Structural design of pits and grates varies with loads and dimensions; contact a knowledgeable engineer for structural design.

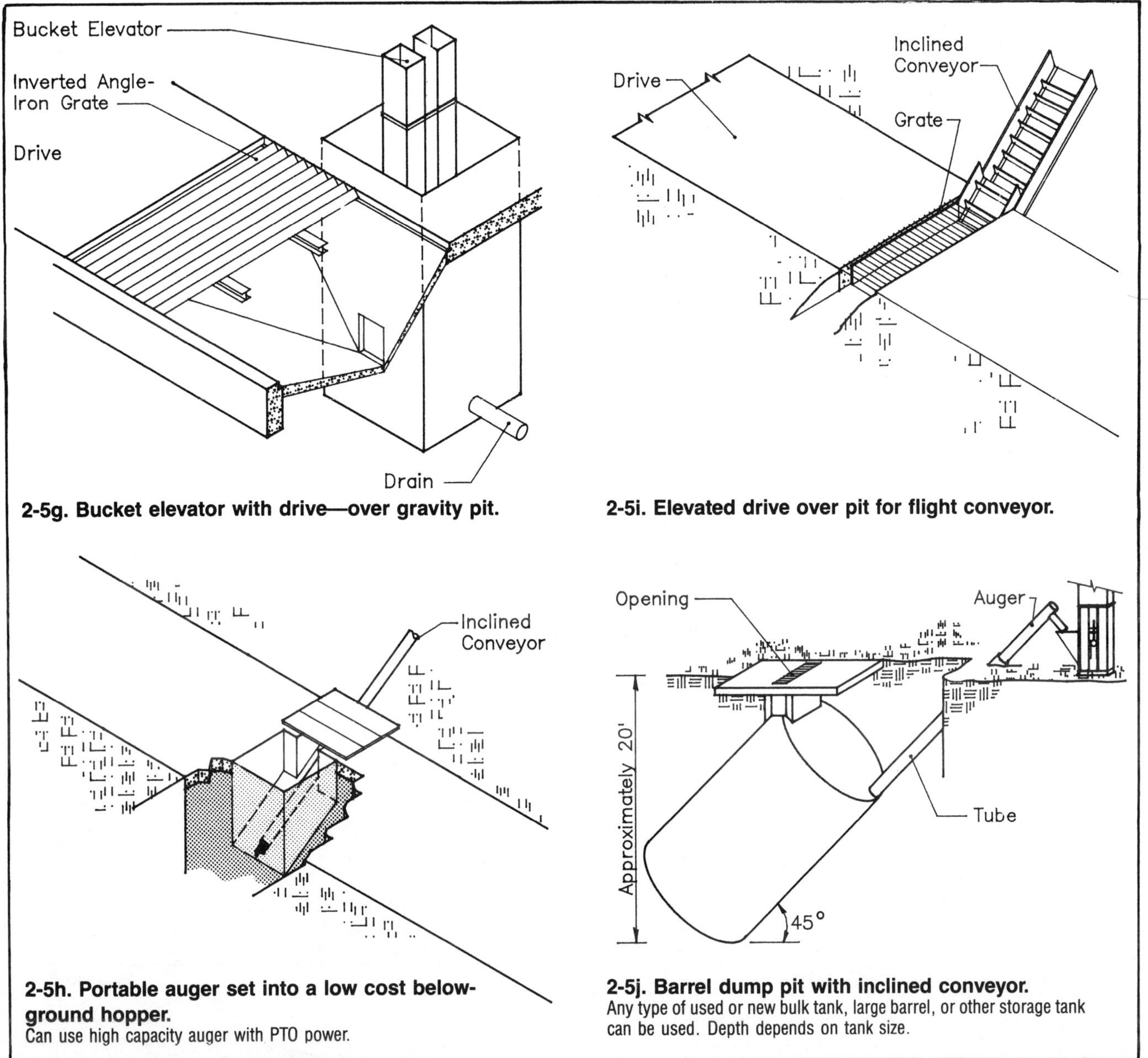

2-5g. Bucket elevator with drive—over gravity pit.

2-5i. Elevated drive over pit for flight conveyor.

2-5h. Portable auger set into a low cost below-ground hopper.
Can use high capacity auger with PTO power.

2-5j. Barrel dump pit with inclined conveyor.
Any type of used or new bulk tank, large barrel, or other storage tank can be used. Depth depends on tank size.

Fig 2-5. Below-ground dump pits continued.

Grain is moved from dump pits, Fig 2-5, to the elevating conveyor by a conveyor or by gravity. A small (less than 50 bu) dump pit can be the hopper for a high capacity auger, flight elevator, or bucket elevator. Small dump pits are usually shallow to reduce ground water problems. A dump pit as big as the largest transport vehicle permits fast unloading and can be a surge bin for a relatively low capacity bucket elevator, auger, or pneumatic conveyor, as well as a wet holding bin for a small dryer. Large dump pits can lead to deep excavation and possible ground water problems.

A cover grate over dump pits is essential for safety and allows driving over the pit. Make the grate area large enough to receive grain without spillage. Plan convenient vehicle access—backing a truck or a wagon to unload is dangerous and inconvenient.

Dump pit augers can overload if too much flighting is exposed to grain; cover the exposed intake flighting, Fig 2-5. Gates, baffles, half-pitch flighting, or auger sections with smaller or tapered flighting also reduce pit auger overload.

Pave at least 8'x8' of the grain receiving area with concrete. Refer to *Farm and Home Concrete,* AED-26, by Midwest Plan Service.

Divert surface runoff around the grain center. Slope the concrete at least ¼"/ft away from the dump pit opening. Keep rain and snow out of the pit with a solid cover. Provide drainage in the bottom of below-ground pits. Protect deep pits from ground water by installing tile around the pit a few inches below the lowest point of the pit. Drain the tile by gravity or sump pump. If your water table is high, build a shallow dump pit on earth fill for better drainage, Fig 2-6.

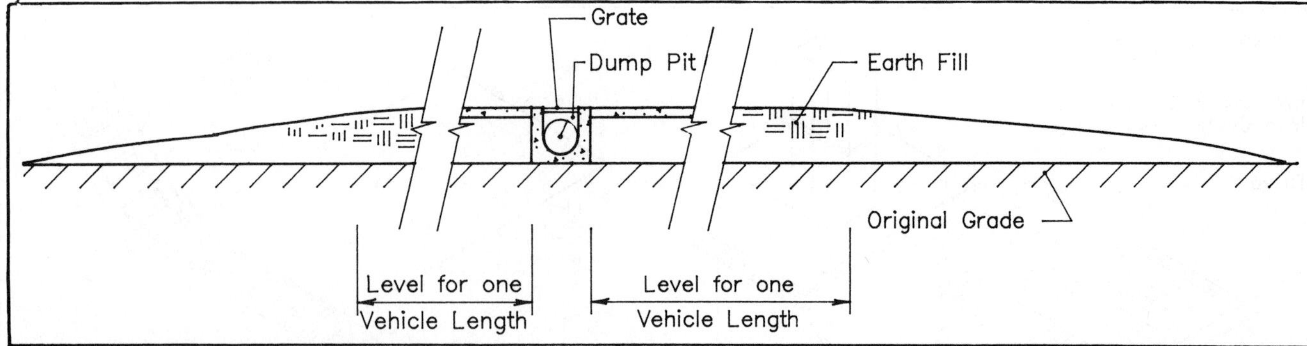

Fig 2-6. Dump pit in earth fill above grade.
Earth fill above grade for dump pit drainage. Build the ramp up to 4′ above grade and at least 16′ wide. Extend the ramp at least one vehicle length in each direction. Consider concrete pavement.

A building over the receiving area gives all-weather protection and convenience, Fig 2-7. This building can house management, control, and feed processing facilities. If necessary, leave space and plan the layout to add a building later. Leave at least 3′ from the door to the edge of the pit to reduce rain and snow blow-in. Recommended doors are 14′x14′ but never less than 12′ wide. Make them 16′ wide for large tractors with dual wheels.

A dump pit near one end of the building lets long trucks park outside to dump grain. For indoor truck hoisting, run the roof trusses parallel to and beside the drive, so a 14′ wide head space over the drive is open for hoisting.

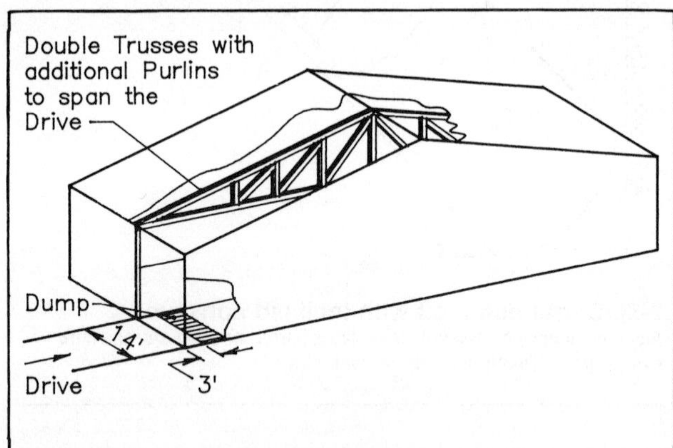

Fig 2-7. Grain receiving building.

Consider installing overhead hopper bottom bins in the receiving area for fast grain loadout and to assist blending grain at marketing. These bins can also hold grain for drying or feed processing. Loadout bins can also be just outside the receiving building near the dump pit or inside the building. On large farms, grain receiving and marketing loadout can conflict at harvest; consider putting overhead loadout bins in a separate lane around the perimeter of the complex. Size loadout bins for at least one semi-truck load (at least 800 bu).

Grain centers with bucket elevators usually combine receiving and loadout because the bucket elevator discharge is directly above the grain receiving area, Fig 2-8. Inclined conveyors tend to separate receiving from loadout because the intake is farther from the discharge, Fig 2-9.

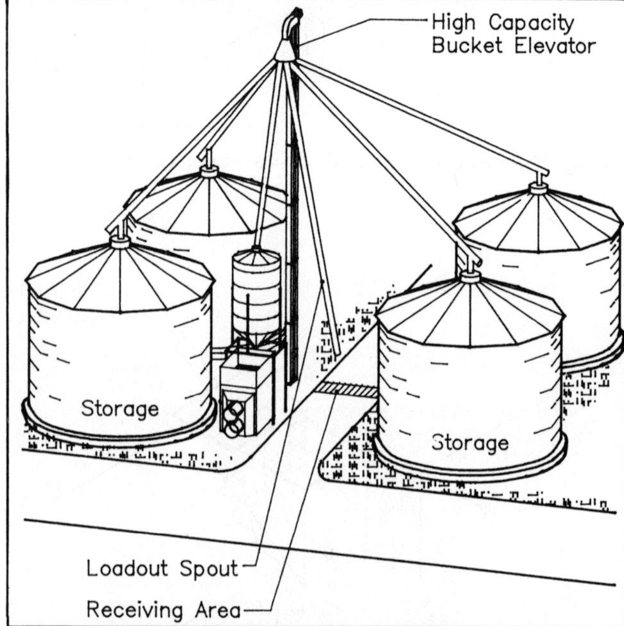

Fig 2-8. Example bucket elevator system.
Receiving pit and loadout spout are in the same area.

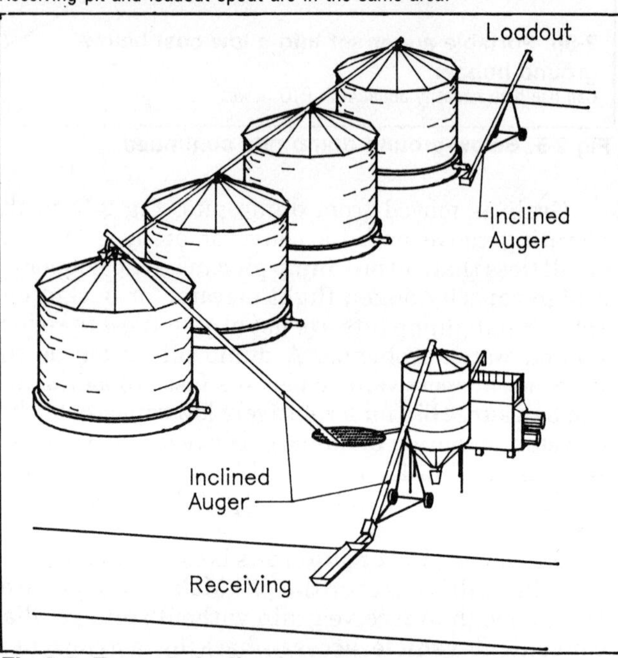

Fig 2-9. Example inclined conveyor system.
Receiving and loadout systems are in separate lanes to avoid conflict. Options to the portable-inclined loadout auger shown include vertical or inclined conveyors in each bin, Figs 2-15 to 2-17.

Grain receiving capacity

With no dump pit, the grain receiving conveyor must have high capacity to keep up with harvesting. With a large dump pit, the conveyor's capacity can be as low as the harvesting rate.

For transport vehicles of different capacities, size grain receiving conveyor capacity on the time to unload the largest vehicle. In a two vehicle system, the largest vehicle must travel to the grain center, unload, and return while the smallest vehicle fills. Estimate grain receiving capacity for a two vehicle system with the following steps. See Fig 2-10 for terms used in the example.

1. Estimate maximum harvest rate (MHR, bu/hr) and determine grain capacity of the largest (LV, bu) and smallest (SV, bu) transport vehicles. See the section on harvesting rate in Chapter 6.
2. Estimate travel time for the largest vehicle to go from the farthest field to the grain center (TTG, min), and to return to the field (TTR, min). Empty vehicles may travel faster than loaded ones. With time in minutes, distance in miles, and speed in miles/hr; calculate estimated travel times:

   $$\text{TTG or TTR} = 60 \times \text{distance} \div \text{speed}.$$

3. Assume at least 5 min for miscellaneous activities (TMA, min) such as checking moisture content, starting conveyors, and cleanup.
4. Calculate time to fill the smallest vehicle (Sfill, min):

   $$\text{Sfill} = \text{SV} \times 60 \div \text{MHR}$$

5. The largest vehicle travels to the grain center, unloads, and returns while the smallest vehicle fills. Calculate time available for the largest vehicle to unload at the receiving area (AUT, min):

   $$\text{AUT} = \text{Sfill} - \text{TTG} - \text{TMA} - \text{TTR}$$

6. Unloading conveyor capacity (UC, bu/hr) can be as low as the harvest rate if a dump pit acts as a surge tank between loads. Calculate the minimum receiving pit capacity required for the unloading capacity to equal the harvest rate (PC_{hr}, bu):

   $$PC_{hr} = LV - (MHR \times AUT \div 60)$$

   If actual pit capacity (PC_a, bu) is greater than PC_{hr}, UC can be as low as the harvest rate. Larger unloading conveyor capacity allows more flexibility for changes in harvesting rate or transport vehicles.

   If PC_a is less than PC_{hr} or if there is no pit, calculate the unloading capacity (UC, bu/hr) ($PC_a = 0$ if no pit):

   $$UC = (LV - PC_a) \times 60 \div AUT$$

Example 2-1:

1. Determine required unloading capacity for a 400 bu/hr (MHR) harvest rate, one 500 bu (LV) transport vehicle, and one 300 bu (SV) transport vehicle.
2. The largest transport vehicle's speed averages 10 mph (both ways) from the farthest field 1 mile from the grain center.

   $$\text{TTG} = \text{TTR} = 60 \text{ min/hr} \times 1 \text{ mile} \div 10 \text{ mph} = 6 \text{ min each way}$$

3. Assume 5 min for miscellaneous activities (TMA).
4. Calculate the time to fill the smallest vehicle:

   $$\text{Sfill} = 300 \text{ bu} \times 60 \text{ min/hr} \div 400 \text{ bu/hr} = 45 \text{ min}$$

5. Calculate time available for the largest vehicle to unload:

   $$\text{AUT} = 45 - 6 - 5 - 6 = 28 \text{ min}$$

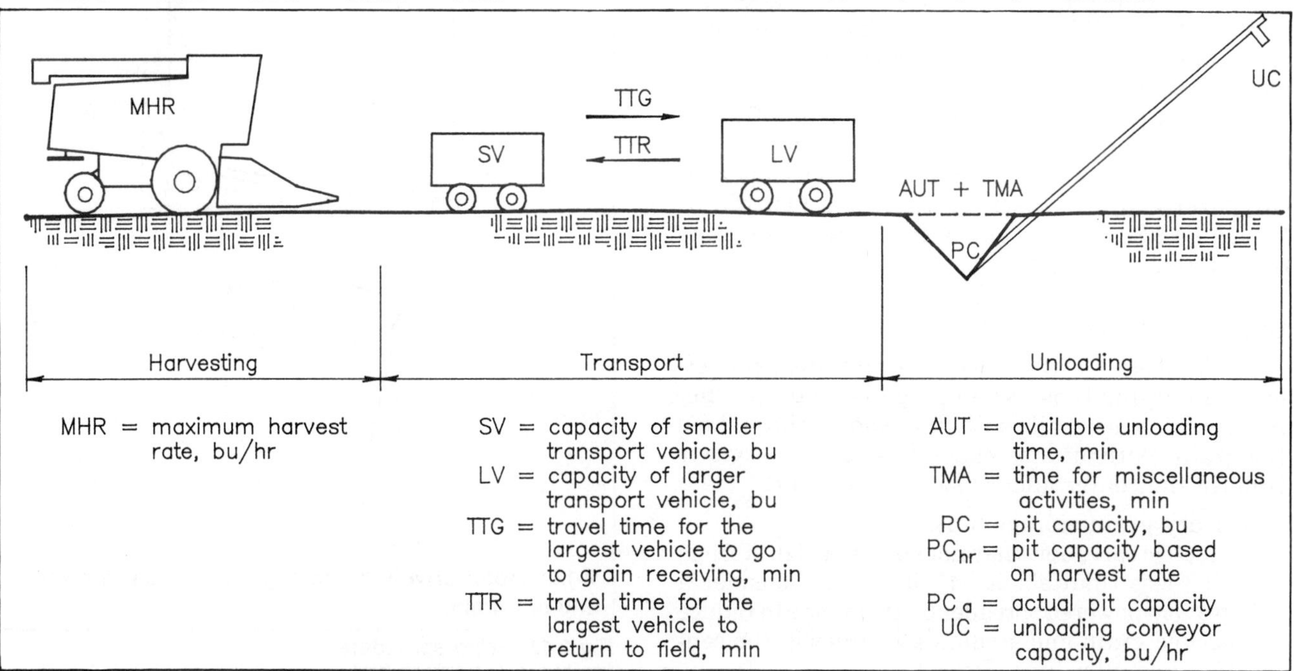

Fig 2-10. Definitions for the example of grain receiving capacity.

6. Calculate minimum receiving pit capacity for unloading to keep up with harvesting rate:

   PC_{hr} = 500 bu − (400 bu/hr × 28 min ÷ 60 min/hr) = 313 bu

   Calculate required unloading conveyor capacity:

   If PC_a is greater than 313 bu, UC can be as low as MHR, i.e. UC = 400 bu/hr.

   IF PC_a is less than 313 bu, e.g. PC_a = 250 bu:

   UC = (500 bu − 250 bu) × 60 min/hr ÷ 28 min = 536 bu/hr

   With no dump pit (PC_a = 0):

   UC = (500 bu − 0 bu) × 60 min/hr ÷ 28 min = 1,071 bu/hr

This example shows that dump pit capacity dramatically influences grain receiving capacity. Consider increasing the unloading conveyor and/or receiving pit capacity to meet future demand. Unloading conveyor and receiving pit capacity may be difficult to change in the future.

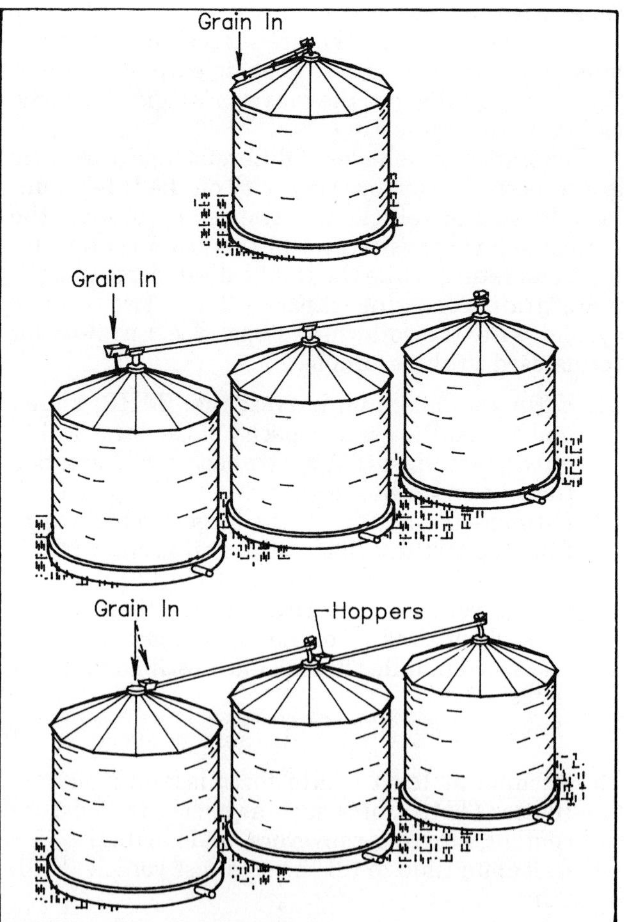

Fig 2-11. Roof conveyor types.

Roof Conveyors

Roof conveyors transfer grain to the top of one or more storages, Fig 2-11. With a bucket elevator, roof conveyors reduce required leg height for gravity spouting, particularly for bins farther than one bin diameter away. Roof conveyors can be difficult to maintain unless catwalks are installed for better access.

Roof conveyors are usually augers, but can be belts or chains. The augers are usually U-troughs, especially for large grain volumes handled frequently. Augers mounted at the bin roof angle commonly fill large diameter bins, because portable conveyors cannot reach the roof peak. Size a roof conveyor for 5%-10% more capacity than the one feeding it. Make sure the feeding conveyor shuts off if the roof conveyor stops.

While a roof auger to more than one bin fills an intermediate bin, some grain can carryover to the end of the auger and plug it. Install drop outs that permit carryover grain to discharge into the last bin or through a spout on the end of the auger to a temporary storage or wagon.

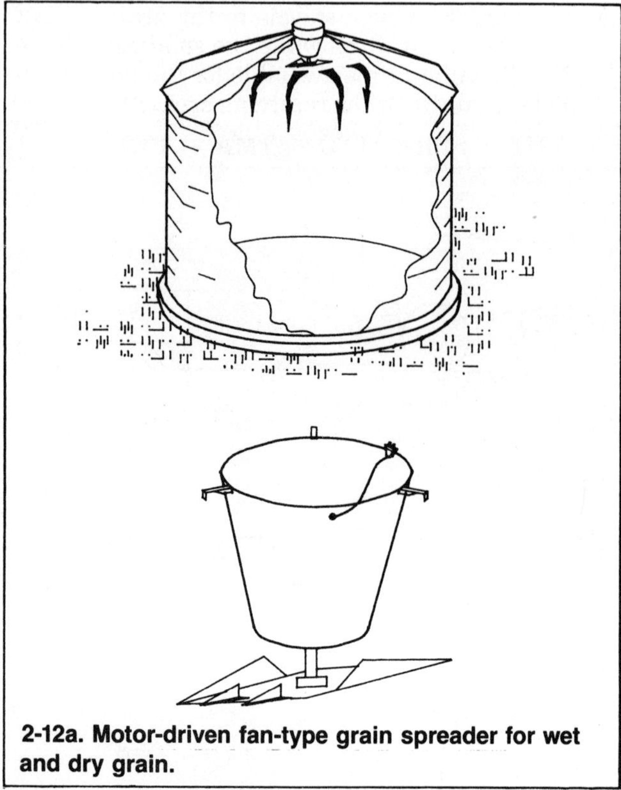
2-12a. Motor-driven fan-type grain spreader for wet and dry grain.

Fig 2-12. Grain spreaders.
Gravity and electrically powered.

Grain Spreaders

Provide electrically-powered grain spreaders (levelers) in drying bins. When properly adjusted, they make a fairly level pile with fines spread throughout the grain. Without a spreader, fines tend to concentrate in the center and cause uneven drying.

For dry grain storage bins:

- Electric spreaders may not be justified in storage bins less than about 24′ diameter. Consider a gravity-powered spreader or grain cone to reduce packed grain and accumulated fines in the center of smaller bins, Fig 2-12.

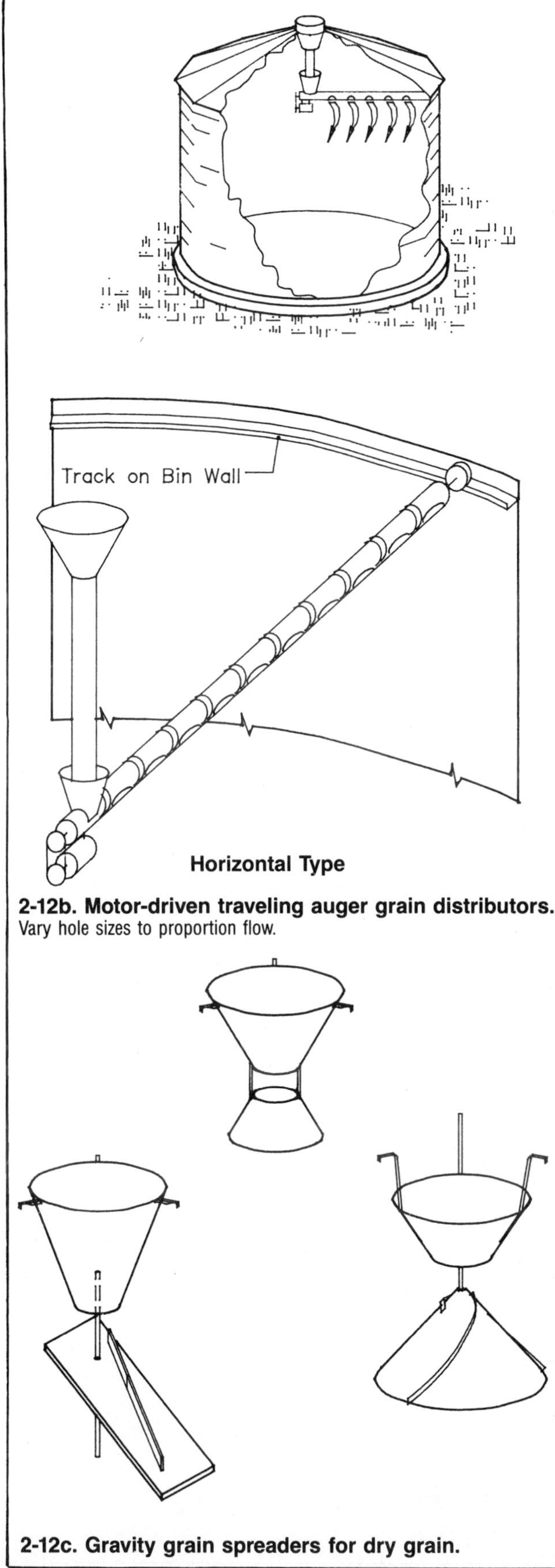

2-12b. Motor-driven traveling auger grain distributors.
Vary hole sizes to proportion flow.

2-12c. Gravity grain spreaders for dry grain.

Fig 2-12. Grain spreaders continued.

- Another option is not to use grain spreading and let fines concentrate in the center. Frequent withdrawal of the center core of grain is needed to remove accumulated fines. This method works best for bins 30′ diameter and larger. Direct the flow of grain toward the center of the storage bin. Remove grain at regular intervals (daily or more often) during filling to remove the peak. Withdraw grain until the center 4′-8′ diameter surface is an inverted cone, Fig 2-13. See AED-20, *Managing Dry Grain in Storage,* for more information.

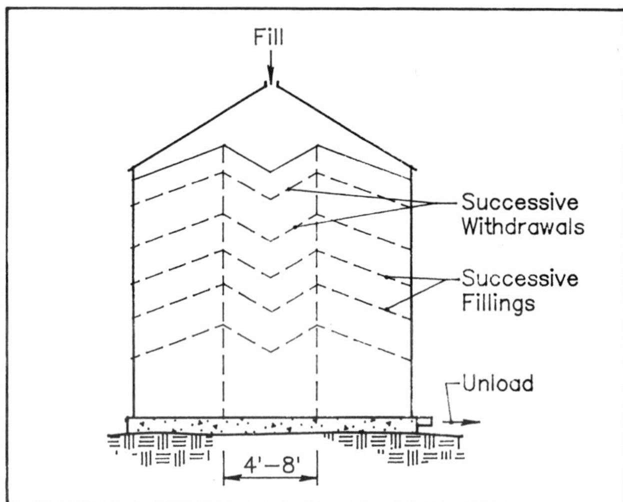

Fig 2-13. Withdrawals to remove fines in storage bins.
Repeated withdrawals remove many of the fines that accumulate in the center core.

Grain Cleaning

Fines, foreign material, and broken kernels are grain handling problems. Kernels break during harvesting and handling. Cleaning grain before storing improves storability. But, unless fines cause serious storage problems, cash grain farmers with no market for the cleanings may lose money by cleaning grain. The broken grain, fines, and foreign material must be collected, perhaps stored, and fed, sold, or discarded.

Select a grain cleaner that collects and conveys screenings away with little or no attention. Some options for locating the grain cleaner in a handling system are shown in Fig 2-14. The most common locations are: at receiving, after the dryer just before delivery to storage, and at loadout.

Cleaning is easier and more complete at low flowrates, e.g. at the output of a continuous flow dryer. Collect screenings in a wagon or a special bin. Dry grain is more effectively cleaned than wet grain and dry screenings are easier to store than wet ones. If grain is dried and stored in the same bin, clean the grain before drying.

Corn screenings are good feed. Feed wet screenings within a day or two after collecting. You can give or sell screenings to local livestock farmers. Take care when feeding fines from damaged grain. Moldy, soft, partially rotted grain particles are screened with the fines and may include toxins harmful to livestock.

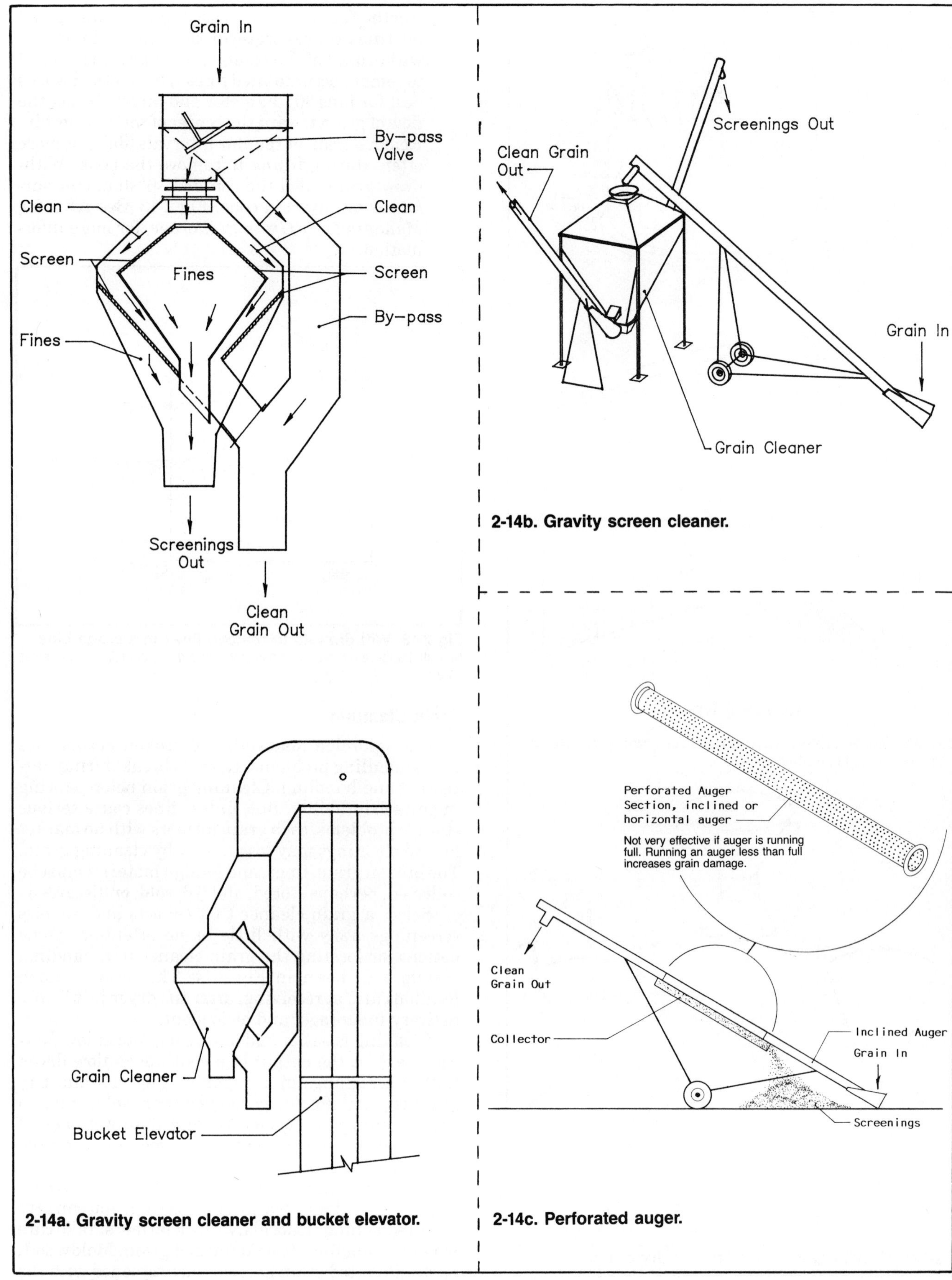

2-14a. Gravity screen cleaner and bucket elevator.

2-14b. Gravity screen cleaner.

2-14c. Perforated auger.

Fig 2-14. Grain cleaners.

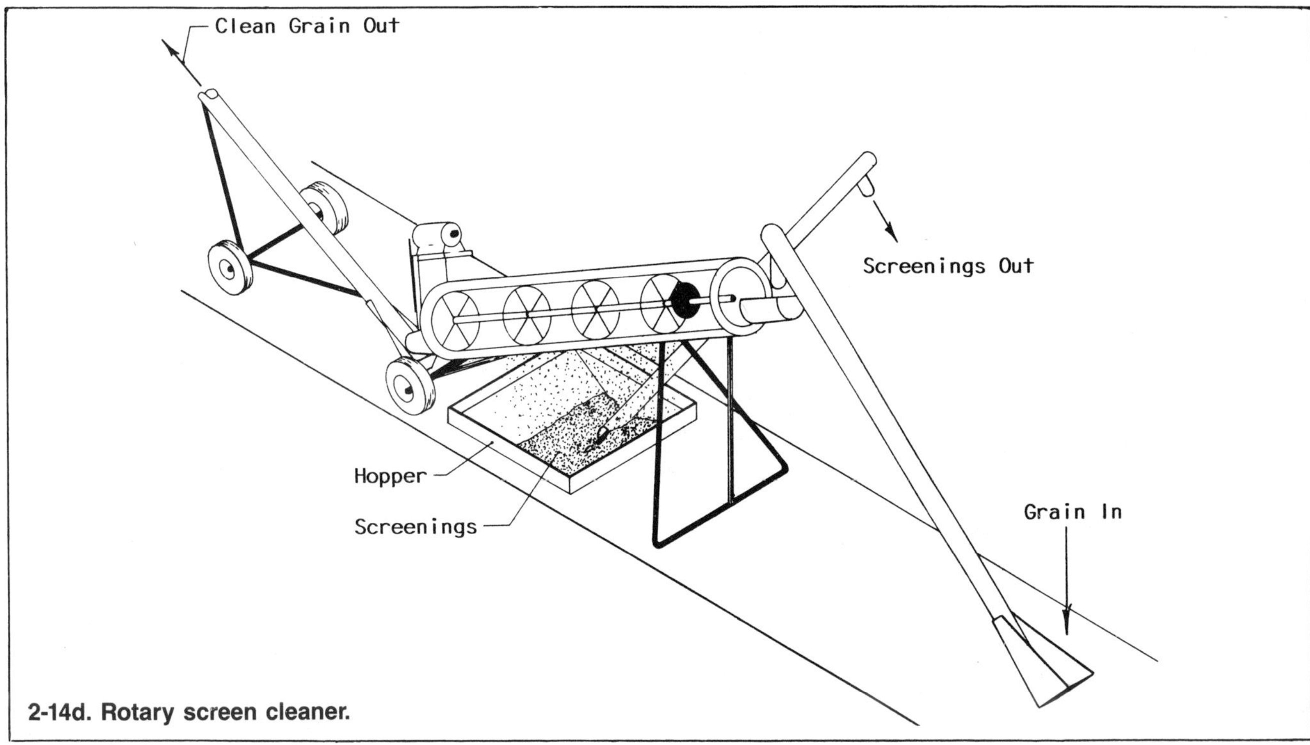

2-14d. Rotary screen cleaner.

Fig 2-14. Grain cleaners continued.

Grain Loadout

Storage bins can unload directly to transport vehicles by:

- A portable inclined auger fed by an underfloor auger, Fig 2-15. Sweep augers remove the grain that does not flow out by gravity.
- An attached vertical auger fed by an underfloor auger, Fig 2-16. Vertical augers convey relatively slowly and require high horsepower, but large sizes can provide adequate capacity. Sweep augers remove grain that does not flow by gravity.
- A permanent inclined auger through the bin wall down to the bin floor center, Fig 2-17. Adequate unloading capacity may require a large size (9″-12″ dia.). Recess the intake into the bin floor so a sweep auger can clean at least ¾ of the bin. The bin floor can be a hopper, but cost and water problems will usually eliminate this op-

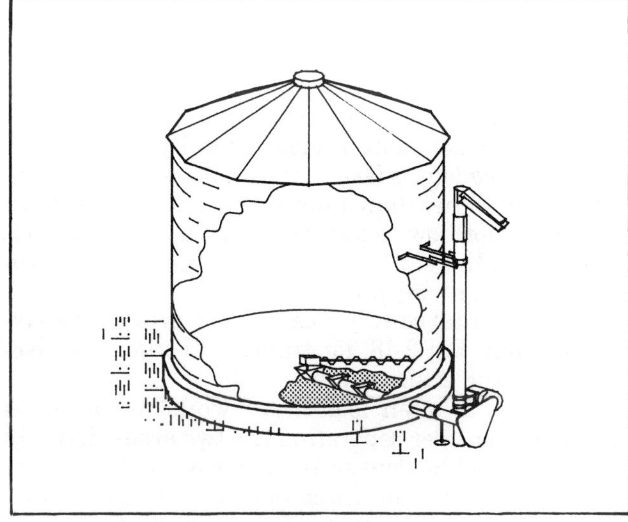

Fig 2-16. Vertical conveyor bin loadout.

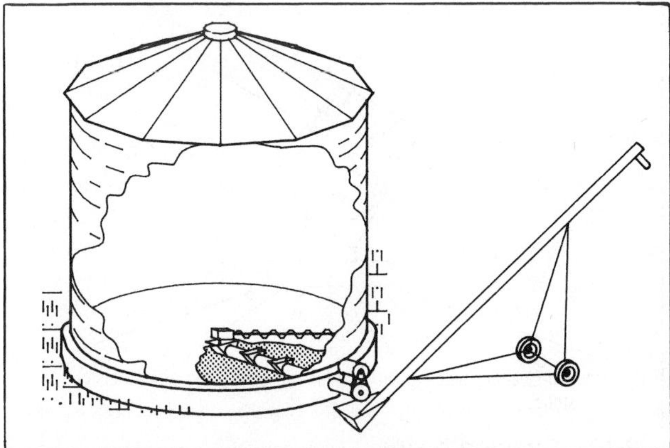

Fig 2-15. Inclined conveyor bin loadout.

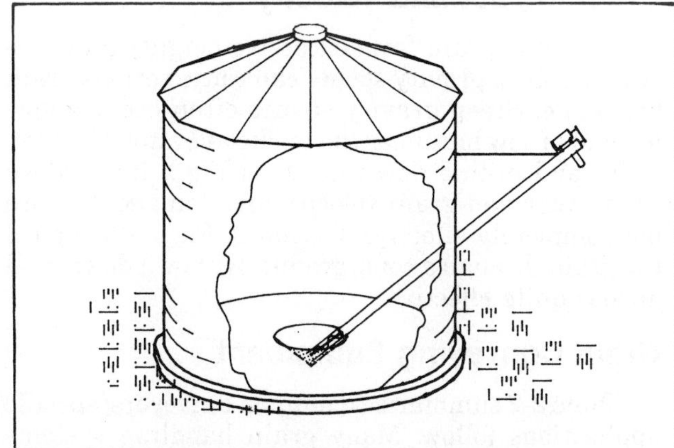

Fig 2-17. Inclined conveyor through bin wall.

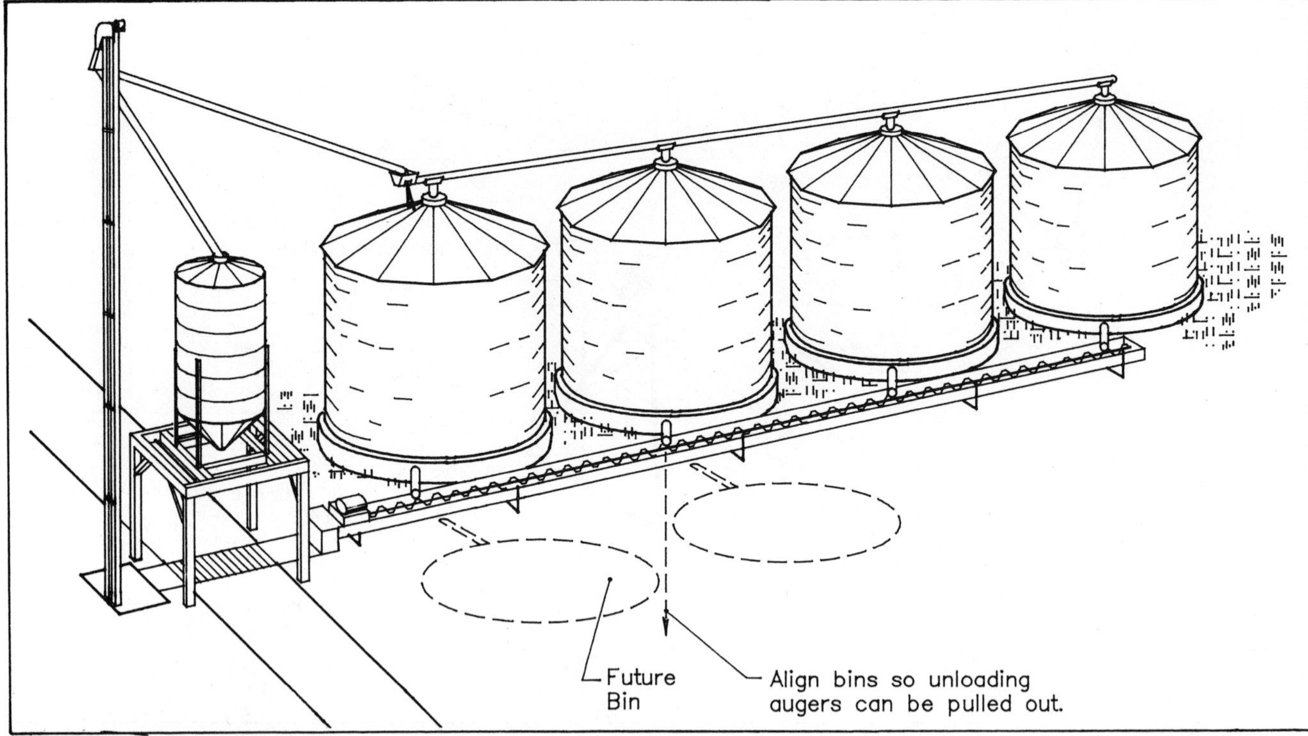

Fig 2-18. Loadout to horizontal collector auger.
Provide for pulling the unloading augers if you add bins on the other side of the collector auger. Set the unloading augers at an angle to the collector auger, or install two-piece unloading augers.

tion. Support the auger near the truck drive with a securely fastened cable. Electrically powered augers require more than 10 hp in this application for adequate loadout capacity.

Grain loadout is faster if grain flows to the intake by gravity rather than a sweep auger. Leave the grain that does not flow to the intake in the bin to form a hopper. At the end of the season, remove the grain hopper with a sweep auger.

A horizontal conveyor can return grain to receiving/loadout, Fig 2-18, for transfer to trucks or elsewhere in the grain center.

If grain is often loaded out while wet grain is received, consider separating the two areas. Usually receiving and loadout in the same drive do not conflict until grain volumes exceed about 100,000 bu/yr.

Grain Impact and Velocity

Dropping grain from excessive heights into bins or down steep gravity spouts can cause serious grain breakage. Steep gravity spouts often occur above loadout or low bins near the bucket elevator. Cushion boxes and in-line flow retarders, Fig 2-19a, reduce spout wear and grain velocity and damage, but are not completely effective. Cyclones, Fig 2-19b, spiral the grain through a cone, gradually slow it down, and appear quite effective.

Grain Conveying Equipment

Table 2-2 summarizes common conveyors; specific applications follow. Many grain handling systems have more than one conveyor type and conveying capacity.

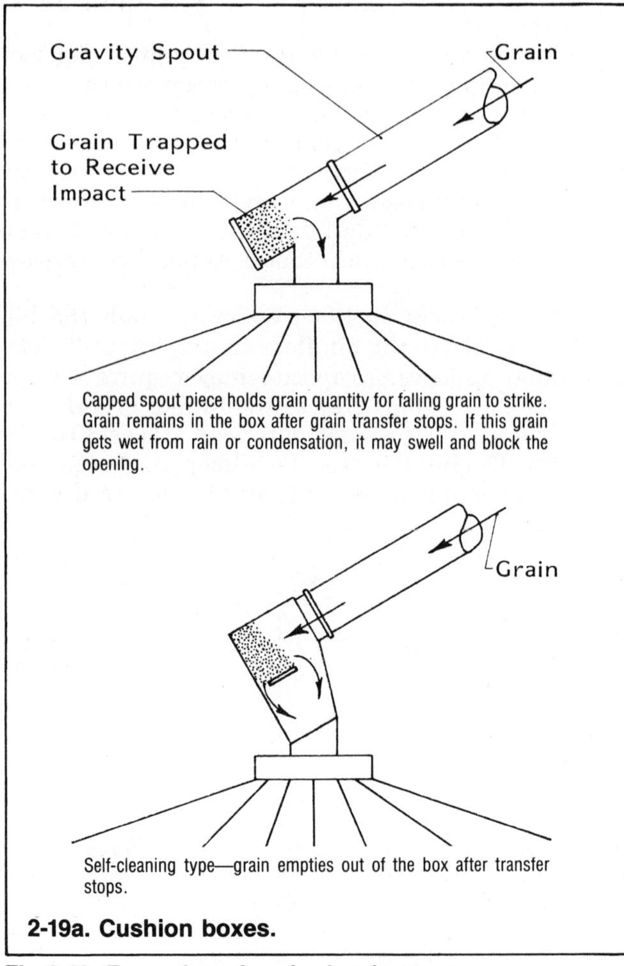

Capped spout piece holds grain quantity for falling grain to strike. Grain remains in the box after grain transfer stops. If this grain gets wet from rain or condensation, it may swell and block the opening.

Self-cleaning type—grain empties out of the box after transfer stops.

2-19a. Cushion boxes.

Fig 2-19. Examples of grain decelerators.

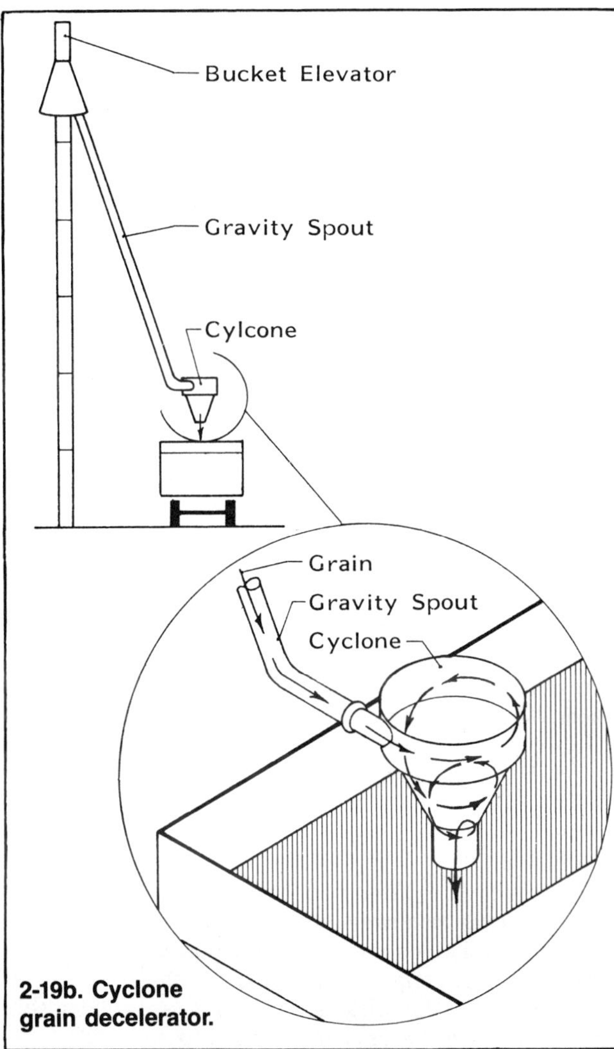

2-19b. Cyclone grain decelerator.

Fig 2-19. Examples of grain decelerators continued.

Auger Conveyors

Augers convey grain at any angle from horizontal to vertical, can be installed or portable, are relatively low cost, and come in many diameters and capacities. Large diameter augers have high capacity but require high horsepower. Compared with other conveyors, augers generally have low capacity per horsepower, especially for wet grains. For example, compared with dry shelled corn, high moisture shelled corn requires nearly twice the horsepower for half the capacity. High capacity farm augers are usually tractor PTO powered.

The simplest tube augers have bearings only at the drive end. When run empty, flighting wear and noise are high. Better tube augers have bearings at both ends and at intermediate points depending on length. Usually they have heavier gauge flighting and casings. Properly spaced bearings prevent the flight from wearing on the enclosure when empty.

Contrary to common belief, augers are not a primary source of grain damage if operated full at their rated speed. Damage can occur when grain is pinched between the auger flighting and casing. Prevent pinching by selecting augers with the clearance between the auger flighting and casing either greater or less than the grain diameter being conveyed.

Table 2-2. Grain conveyors.

Type of conveyor	Horsepower requirement per unit capacity	Advantages	Disadvantages
Auger	Low to medium with dry grain and medium to high with wet grain.	Simple, widely available in many sizes. Low cost. Available for horizontal, inclined, vertical applications; portable, wheeled, or fixed. Adapted to most grain and feed materials. Useful as mixer, flow meter, force feeder, or agitator.	High torque and power required in wet grain. Medium to heavy wear. Noisy if not bearing-supported and/or not operated at rated capacity. High grain damage if not operated at rated capacity.
Belt	Low	Good for long distances. Low power requirement. Quiet. Least handling damage. Capacity only affected by grain weight. Self cleaning.	Limited in angle of elevation. Expensive. Belt maintenance.
Bulk or mass flow.	Low	Good for long distances. Low power requirement. Quiet. Low maintenance. Little grain damage.	Expensive
Flight	Low to medium	Low cost. Flexible. Useful for materials other than grain.	Noisy. Heavy wear.
Bucket	Low	Efficient, compact. Low maintenance. Quiet. High capacity for vertical lift. Reliable and adaptable to automation. Easily cleaned.	Difficult to erect and change (increase capacity). Expensive. Grain damage high for large drop heights. Elevator head service is difficult.
Pneumatic	High	Flexible installation. Easily cleaned. Convenient grain delivery to many locations.	High power requirement. Creates dust, usually requires separation equipment.

Maximum recommended auger speeds are:

- 800 revolutions per minute (rpm) for 4″ augers.
- 600 rpm for 6″.
- 450 rpm for 8″.

Consider a larger auger and run it slower to decrease wear, increase power efficiency, and reduce grain damage. For example, a continuous flow dryer has a typical output of 200 to 400 bu/hr. Choose a 6″ or 8″ auger rather than a 4″ one, and run it slower. Slow speeds may require a gear reducer, which is a good investment over the life of the equipment. Augers are not generally used for handling edible beans or seed grain because they can damage the seed coat.

U-trough augers are preferred for horizontal conveying of large volumes. Consider them for hard to maintain locations, such as overhead augers and augers in dump pits. They operate at slower speeds, which increases auger life and may reduce grain damage, but are more expensive. U-trough augers do not work well at inclines over 15° because the enclosure does not hold grain in the flights. The top cover is removable for repair. U-trough augers may require more support than tube augers. Catwalk support systems are common for overhead installations.

Capacities and power requirements

Auger capacity and horsepower are affected by diameter, pitch, speed (rpm), exposed intake length, incline, and grain properties. Auger length affects power requirement, but not capacity.

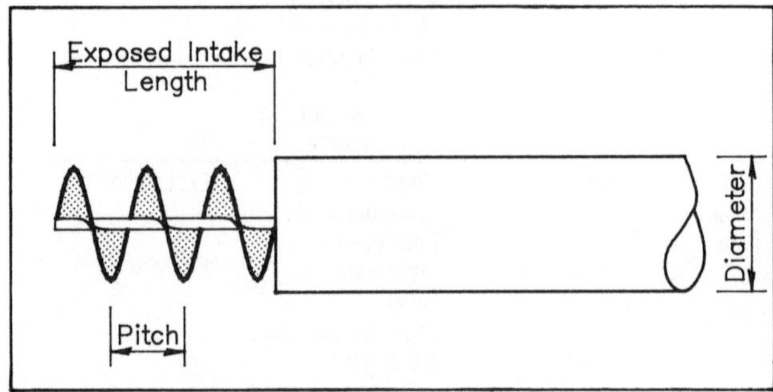

Fig 2-20. Auger definitions.

Auger capacity increases with **diameter.** A **pitch** length equal to flight diameter (single pitch) gives the best performance under most conditions. Most farm augers are single pitch.

Capacity and required horsepower increase with **speed.** The increase is less for an inclined auger or short intake length. Capacity increases until the grain has difficulty entering the auger because the intake flighting throws it back out—usually at speeds beyond normal operation. Grain damage and auger wear tend to increase with speed, especially if the auger is not run full.

Capacity and power requirements increase with **exposed intake length.** At some point, more exposed intake increases horsepower but not capacity. Optimum exposed intake length at full capacity and rated rpm is generally 2 to 3 times the auger diameter.

For a given horsepower, increased auger **incline** decreases capacity. Horsepower increases from horizontal to a maximum at inclines between 45° and 60°. Above a 60° incline, capacity loss reduces horsepower. At 90°, the power required is about the same as for horizontal, but capacity is much less.

Higher **grain moisture content** reduces capacity and increases power requirements because wet grain is heavier and has more friction.

Grain type influences capacity and power requirements. An auger's capacity for wheat, grain sorghum, oats, barley, and rye is a little less than for corn. With soybeans, an auger has lower capacity and higher power requirements than with corn.

Substantial **grain depth** over the auger intake, as in bulk tanks, increases auger horsepower needs, especially with long exposed intakes. Half pitch flights (pitch = ½ auger diameter) run at rated rpm can reduce the amount of grain entering the auger and increase leverage, so horsepower is reduced.

Table 2-3 gives capacities and power requirements of single pitch augers handling dry shelled corn. Adjust Table 2-3 for different auger speeds with Table 2-4 and for wet corn with Table 2-5.

Sizing motors, pulleys, and V-belts is discussed in Chapter 6. The values from Table 2-3 do not include drive-train losses—increase by 10% to determine required motor power. If an auger occasionally handles wet grain, size the motor for wet grain power requirements.

Table 2-3. Estimated auger capacity and power.

For dry (14% maximum moisture content) corn. Values for 4″ and 6″ augers are based on a lot of data. Values for 8″ and 10″ augers are based on very limited data. Values for 12″-16″ augers were extrapolated. Actual auger performance may vary so use manufacturer's data for designing auger systems. Multiply dry corn values by 0.6 for wet corn capacity. Use the table values for wheat, grain sorghum, oats, barley, and rye because actual values are only slightly less. For soybeans, multiply capacity by 0.75 and power by 1.10. Multiply bu/hr by 50 for approximate lb/hr capacity for meal or concentrate feed.

Auger dia., in.	Auger speed[1] rpm	Incline angle 0° bu/hr	0° hp/10′	25° bu/hr	25° hp/10′	35° bu/hr	35° hp/10′	45° bu/hr	45° hp/10′	90° bu/hr	90° hp/10′
4	900 [2]	560	0.6	500	0.9	480	0.9	450	1.0	270	0.8
6	600	1,500	1.0	1,350	1.5	1,290	1.6	1,190	1.6	710	1.3
8	450	2,210	1.4	1,990	2.2	1,890	2.2	1,760	2.3	1,050	1.8
10	360	3,300	2.0	2,970	3.1	2,830	3.2	2,620	3.2	1,570	2.5
12	300	4,520	2.5	4,070	3.9	3,870	4.0	3,590	4.0	2,150	3.2
14	260	6,230	3.4	5,610	5.3	5,340	5.4	4,950	5.5	2,960	4.3
16	225	8,040	4.4	7,240	6.8	6,870	7.0	6,390	7.1	3,820	5.6

[1]Auger speeds for 3,600 in/min flighting velocity along auger length (theoretical grain velocity) for all diameters.
[2]4″ auger at 900 rpm vibrates excessively. 900 rpm values are for converting with Tables 2-4 and 2-5.

Table 2-4. Conversions for auger speed.
For converting Table 2-3 for different auger speeds. You can interpolate between speeds with reasonable accuracy. These values are for dry corn—convert first for speed, then for moisture content, Table 2-5.

Speed relative to Table 2-3 %	Multiplier Capacity	Power
125	1.17	1.23
100	1.00	1.00
75	0.79	0.76
50	0.56	0.51
25	0.29	0.26

Table 2-5. Conversions for wet (25%) corn.
Wet corn **capacity** = 0.6 × dry corn capacity (Table 2-3). This table converts power requirements from Table 2-3 for wet corn. If the auger also operates at a different speed, first convert for speed, Table 2-4, then for moisture.

Speed relative to Table 2-3 %	Incline angle 0°	25°	35°	45°	90°
		- - -hp multiplier - -			
125	2.6	1.9	1.8	1.7	1.0
100	2.8	2.0	1.9	1.8	1.0
75	3.1	2.2	2.1	2.0	1.0
50	3.8	2.7	2.6	2.5	1.0
25	5.4	3.8	3.5	3.3	1.0

Example 2-2:

What is the estimated capacity and required motor size for a 51′ long, 6″ diameter auger operating at 450 rpm with a 35° incline?

Solution:

From Table 2-3, a 6″ auger running at 600 rpm and a 35° incline, conveys 1,290 bu/hr of dry corn and requires 1.6 hp per 10′ of auger length.

Determine relative operating speed:

(450 rpm ÷ 600 rpm) × 100 = 75%

From Table 2-4, relative capacity multiplier is 0.79 and relative power multiplier is 0.76 when the relative speed is 75%. Calculate the relative capacity and power requirement.

Capacity at 450 rpm = 0.79 × 1,290 bu/hr = 1,019 bu/hr

Horsepower per 10′ at 450 rpm = 0.76 × 1.6 hp/10′ = 1.22 hp/10′

Total hp = (51′ ÷ 10′) × 1.22 hp/10′ = 6.22 hp.

From Table 6-11, select a 7½ hp electric motor or a 12 hp gasoline engine.

Example 2-3:

Determine the estimated capacity and hp for conveying wet corn with the auger of Example 2-2.

Solution:

From Example 2-2, the adjusted auger capacity is 1,019 bu/hr and total conveying hp is 6.22 hp. Adjust the capacity and required power for moisture with Table 2-5. Multiply capacity by 0.6. From Table 2-5, hp multiplier is 2.1 when the relative speed is 75%.

Capacity for wet corn at 450 rpm = 1,019 bu/hr × 0.6 = 611 bu/hr.

Total hp for wet corn at 450 rpm = 6.22 hp × 2.1 = 13.06 hp.

From Table 6-12, select a 15 hp electric motor or 23 hp gasoline engine.

Example 2-4:

Determine the capacity and hp if conveying soybeans with the auger in Example 2-2.

Solution:

From Example 2-2, the adjusted auger capacity is 1,019 bu/hr and total conveying hp is 6.22 hp. Adjust the capacity and power values for soybeans by multiplying capacity by 0.75 and hp by 1.10, from Table 2-3.

Capacity for soybeans at 450 rpm = 0.75 × 1,019 bu/hr = 764 bu/hr.

Total hp for soybeans at 450 rpm = 1.10 × 6.22 hp = 6.84 hp.

From Table 6-12, select a 7½ hp electric motor or a 12 hp gasoline engine.

Ease of maintenance is important. Bin unloading augers must be removed for repair or maintenance. Locate them with plenty of room to pull the auger completely out of the bin without cutting the auger. Place catwalks beside overhead augers for easier and safer access. Covered U-trough augers are easier to maintain than tube augers.

Augers are a major farm hazard. Manufacturers put safety shields on exposed auger drive assemblies and grates over auger intakes to keep hands, feet, and clothing from contacting the auger flighting or being caught between the flighting and housing. Do not remove safety shields. When children are around, **never** start **any** machinery without being sure they are not in danger.

Belt Conveyors

Belt conveyors handle grain gently, but are expensive and generally unsuitable for steep inclines. They have higher capacity per horsepower than augers (i.e. they are more power efficient).

Belt conveyors are not common on Midwest grain farms. They may become more common as facilities get larger and grain damage affects marketing, especially with delicate speciality crops. Farm belt conveyors are used for edible beans and soybean seed.

Bulk or Mass Flow Conveyors

Bulk flow conveyor housings are usually rectangular, but U-trough or round tube designs are also available, Fig 2-22. The flights are usually 8″-12″ wide and often only ¾″-1½″ high. These conveyors are called bulk flow because the shallow flight scraping along the housing bottom carries grain 3 to 5 times as deep as the flight. The grain between the flights pushes along the grain above. The flights are usually high density plastic or composition material pre-

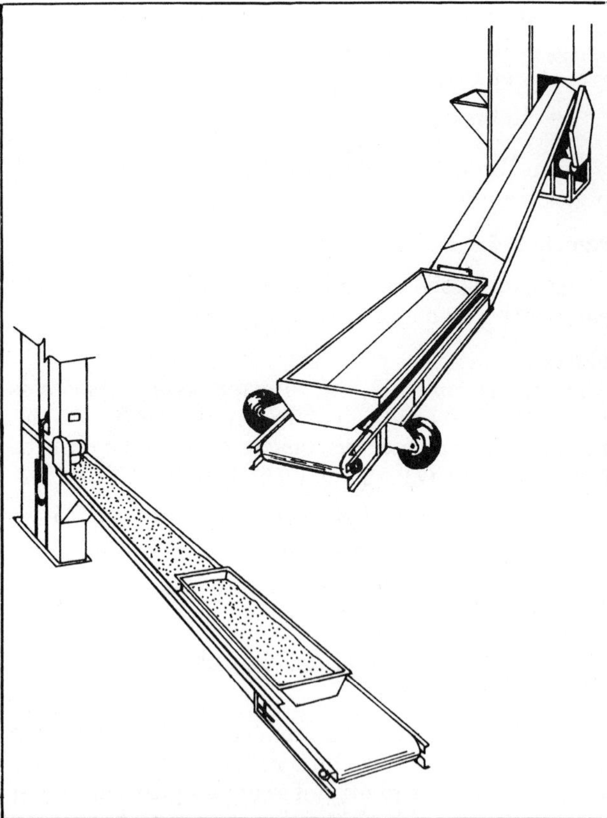

Fig 2-21. Typical farm belt conveyors.

Table 2-6. Typical horizontal belt conveyor capacity.
Belt speed = 100 fpm (feet per minute); uniformly loaded. Capacity depends on belt speed. Capacity at your belt speed = table value × your belt speed ÷ 100. Source: University of California, Agricultural Circular 517.

Belt width in.	Troughed belt bu/hr at 100 fpm belt speed	Flat belt bu/hr at 100 fpm belt speed	Maximum belt speed fpm
12	370	183	350
14	508	228	400
16	675	305	450
18	833	373	450
20	1,074	482	450
24	1,575	707	600

Table 2-7. Typical bulk flow conveyor capacity and power.
For horizontal conveyors only. Chain speed = 200 fpm. Conveying depth equals width and available conveying area is 85% full. From Buhler Conveyors, Hutchinson Corp, and Berreau literature.

Conveyor width and depth, in.	Capacity bu/hr	Power required hp/10′
8	3,840	1.0
9	4,860	1.2
10	6,000	1.4
11	7,260	1.6
12	8,640	1.8
13	10,140	2.1

cisely fit to the conveyor bottom and sides for excellent conveyor cleanout.

Bulk flow conveyors are common in commercial elevators and for seed handling and are becoming more common on farms. They convey across the top or under the bottom of storages and occasionally elevator pits. They are usually more expensive, but very power efficient (high capacity per horsepower) and reliable.

Rectangular conveyors, Fig 2-22a, are used mainly for inclines up to 15°, although some have shallow horseshoe-shaped flights for vertical conveying. Special tubular designs, Fig 2-22b, have paddles filling most of the tube for conveying at inclines up to vertical.

Avoid bulk flow conveyors for hot grain from a dryer in cold weather because moisture condenses in the conveyor. Permanent conveyors exposed to the weather must be raintight to keep water out of equipment and storages.

See Table 2-7 for capacity and power for horizontal bulk flow conveyors at 200 fpm chain speed. To adjust capacity and horsepower requirements for other depths and speeds use:

- Capacity adjustment for conveyor depth:
  Bu/hr = (bu/hr for conveyor width from Table 2-7) × (available conveyor depth ÷ conveyor width)
- Capacity adjustment for chain speed:
  Bu/hr = (capacity, bu/hr from Table 2-7) × (chain speed, fpm ÷ 200 fpm)
- Horsepower adjustment for depth and chain speed:
  Hp/10′ = (hp/10′ from Table 2-7) × (adjusted bu/hr) ÷ (bu/hr from Table 2-7)

Example 2-5:

Determine the capacity of a conveyor that is 9″ wide and 6″ deep.

Solution:

From Table 2-7, conveyor capacity is 4,860 bu/hr for width = depth = 9″ and chain speed = 200 fpm. To convert conveyor depth to 6″:

Bu/hr = 4,860 bu/hr × 6″ ÷ 9″
= 3,240 bu/hr

Example 2-6:

Adjust the conveyor from Example 2-5 for 150 fpm chain speed.

Solution:

From Example 2-5, capacity is 3,240 bu/hr at 200 fpm. Convert to 150 fpm chain speed:

Bu/hr = 3,240 bu/hr × 150 fpm ÷ 200 fpm = 2,430 bu/hr.

Example 2-7:

What horsepower is required for the conveyor in Example 2-6?

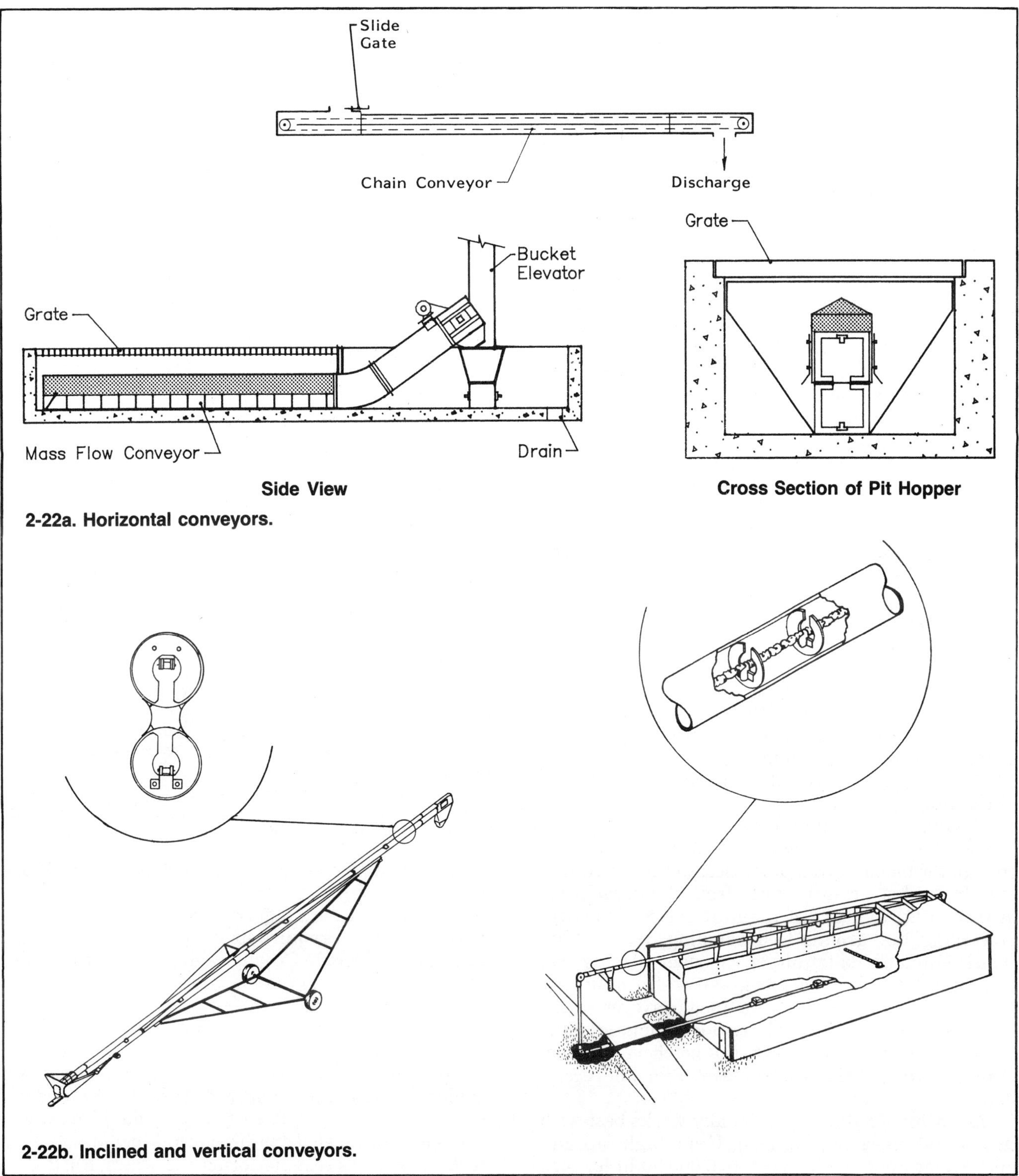

2-22a. Horizontal conveyors.

2-22b. Inclined and vertical conveyors.

Fig 2-22. Bulk or mass flow conveyors.

Solution:

From Example 2-6, adjusted capacity is 2,430 bu/hr. From Table 2-7, capacity is 4,860 bu/hr and power required is 1.2 hp/10′ for the original conveyor. Adjusted power requirements are:

Hp/10′ = 1.2 hp/10′ × 2,430 bu/hr ÷ 4,860 bu/hr
= 0.6 hp/10′

Flight Conveyors

Double chain flight conveyors, like those for ear corn or baled hay, are low cost. Capacity decreases as flight depth or speed decreases and as flight spacing or conveyor incline increase. Although not designed for shelled corn, they can convey wet corn to a holding or drying bin.

The flights are usually 17"-21" wide, 2½"-3" high, and have speeds of 150-200 fpm. The conveyors are typically 25′-60′ long. Shelled corn and small grain capacity is about 1,000 to 2,000 bu/hr at a 15°-35° incline. Motors are 3 to 7½ hp.

Research is not available on grain damage in flight conveyors, but it is believed to be low with flights fitting well in the conveyor housing. A worn conveyor may seriously damage dry grain.

Bucket Elevators

Vertical bucket elevators (legs) are often the hub of convenient, high capacity grain centers. They elevate almost all the grain; other conveyors (usually augers) are used to convey grain to and from the bucket elevator.

Bucket elevators handle wet and dry grain, whole and ground grain, meals, and complete feeds. Capacities and power requirements change little with grain moisture content. Power requirements are low compared with other conveyors, particularly for wet grain. They are quiet, reliable, have little wear, and have a long service life.

A single bucket elevator often does many conveying jobs in one system, Fig 2-23. Occasionally, two functions conflict. For example, transferring wet grain from receiving to wet holding and transferring grain from a continuous flow dryer to storage at the same time. A surge bin can collect dry grain while the bucket elevator is conveying wet grain, or a second elevator or other conveyor can be installed to handle the two functions at once. With one bucket elevator for multiple functions, consider a warning light or electrical lockout to remind or prevent the operator from routing grain into the wrong route, e.g. wet grain into dry storage.

Drying method may determine if one or two bucket elevators are needed. One elevator is usually enough for batch drying processes: wet grain receiving, dry or hot grain transfer from the dryer if required, and dry grain loadout for marketing, processing, and ground feed handling. With a high speed automatic batch dryer, consider placing wet holding over the dryer for fast gravity refill, or lift wet grain to the dryer with an inclined conveyor so the bucket elevator control can be set for automatic transfer of dry or hot grain when the batch is finished. Manually control the bucket elevator for wet receiving.

A continuous flow dryer usually works best with two bucket elevators, Fig 2-24. Use a high capacity bucket elevator (usually 3,000 to 6,000 bu/hr for corn receiving) for wet receiving, for dryer refill if there is no overhead gravity-fill wet holding, and for dryer loadout for marketing or processing. The second, lower capacity, elevator is usually matched to the maximum dryer output (typically 500 to 1,500 bu/hr). It conveys dry grain to final storage or hot grain to bin cooling or dryeration bins. With dryeration, the higher capacity elevator may be taller and spouted to all bins for wet receiving, dryeration transfer, and dry load out. The other is spouted only to the dryeration bins. Consider a gravity surge bin or separate conveyor to transfer wet grain to the dryer to avoid conflict.

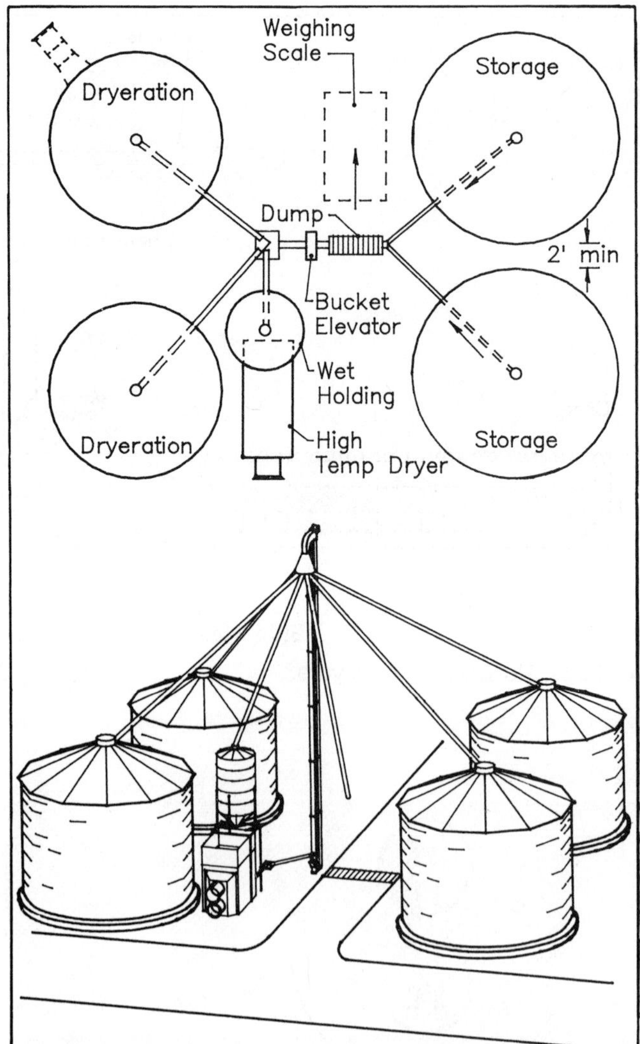

Fig 2-23. Grain center with single bucket elevator.

Bucket elevators require little area, but the support system can require more space, block passage, and restrict growth. Guying cables require more area than tower support systems, Fig 2-25. Typically, guying cables are located at 90° angles to each other. Consult the bucket elevator supplier for required number and location of guying cables. Towers around the elevator can be used to support it in place of guy cables. Towers take up only 8′x8′ or 10′x10′ and reduce obstructions, but cost more. Towers or cages around a bucket elevator also permit including safety features such as rest platforms and stairs, Fig 2-25b.

Nearly all farm bucket elevators are centrifugal discharge, double-leg, with buckets mounted on a belt, Fig 2-26. Material discharges by centrifugal action as the buckets pass over the head wheel. Use slow speed, gravity discharge for seeds and edible beans to reduce impact damage.

Intake hoppers can be on the up leg or the down leg of the bucket elevator. Input to a hopper can be a gravity drop from a pit, spout, or conveyor. Or, an auger can force feed grain directly into the leg at a very low position, Fig 2-27.

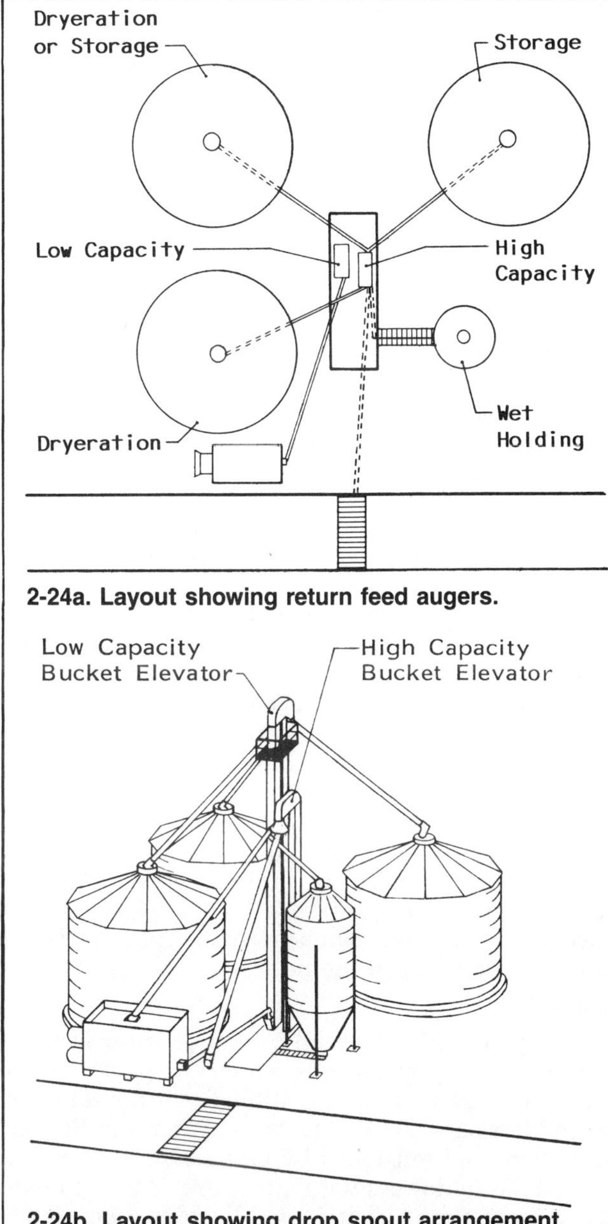

2-24a. Layout showing return feed augers.

2-24b. Layout showing drop spout arrangement.

Fig 2-24. Grain center with two bucket elevators.
Two bucket elevator setup common with a continuous flow dryer. High capacity elevator receives wet grain, discharges to dryer or wet holding, moves grain from wet holding to dryer, and discharges for loadout. Low capacity elevator moves dried grain to dryeration or storage.

Capacities

Bucket elevator capacity can exceed 30,000 bu/hr. Common farm capacities are 1,000 to 6,000 bu/hr, Table 2-8. For a single bucket elevator without a dump pit, at least 3,000 bu/hr is suggested. If the wet grain receiving pit is about as big as the largest grain hauling vehicle, less capacity can work. Table 2-9 gives gravity spout capacities. Table 2-10 gives common bucket elevator applications.

When buying a bucket elevator, consider that elevator capacity can often be doubled with only a 20% cost increase. Elevators are relatively easy and less expensive to build taller when erected, but are difficult and expensive to make taller later. Money spent on greater capacity and discharge height may help

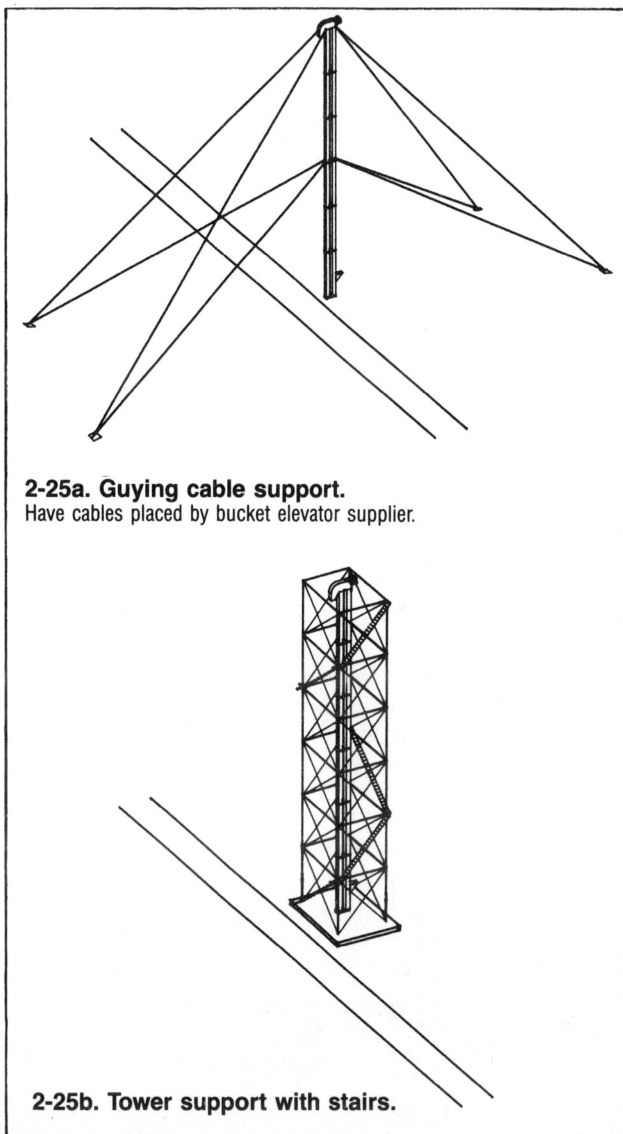

2-25a. Guying cable support.
Have cables placed by bucket elevator supplier.

2-25b. Tower support with stairs.

Fig 2-25. Bucket elevator support systems.

Table 2-8. Typical bucket elevator capacity and power.

Bucket size in.	Bucket spacing in.	Belt speed ft/min	Capacity bu/hr	Power requirements hp/10′ height
4x3	8	240	200	0.10
	6	270	300	0.125
6x4	4¼	270	550	0.20
	4¼	335	700	0.25
7x5	8	335	900	0.30
	6	335	1,200	0.33
9x5	7	265	1,600	0.5
	6	300	1,800	0.5
9x6	12	385	1,500	0.625
	6	385	3,000	1.25
12x7	10	565	5,000	2
15x7	9	565	7,500	3
14x8	10	650	10,000	4

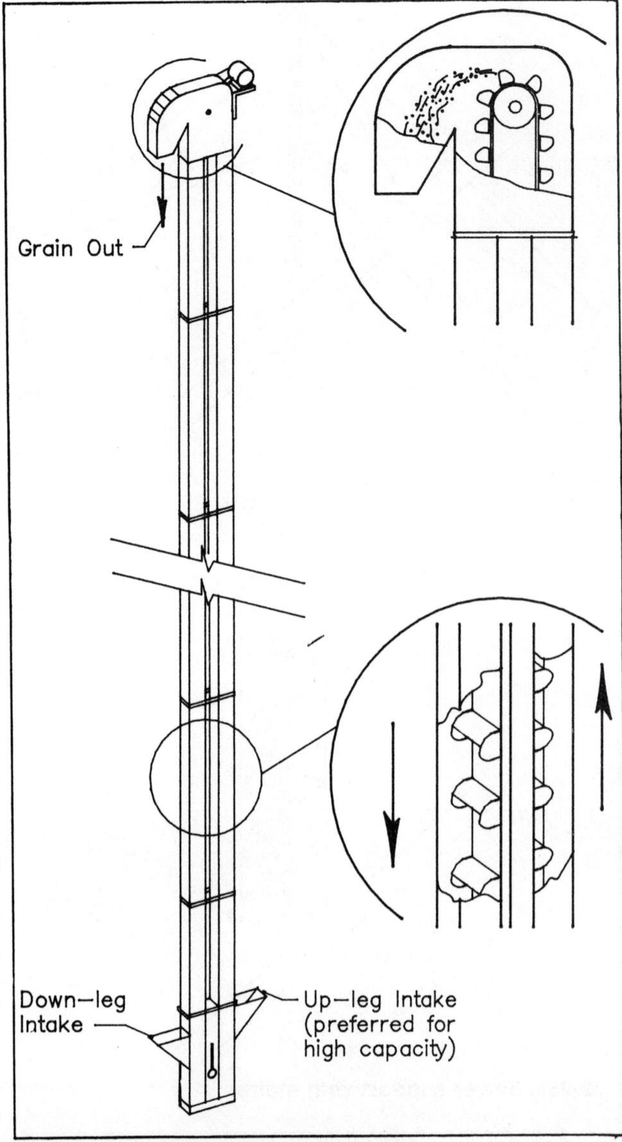

Fig 2-26. Bucket elevator schematic.

Table 2-9. Estimated capacity of down spouts for dry grain.
For 45° slope and generally unrestricted flow. Based on an industry recommendation of 65 bu/hr/in² of downspout open area as a conservative estimate.

Downspout dia., in	Capacity bu/hr
4	800
6	1,800
8	3,300
10	5,100
12	7,400
14	10,000
16	13,100

future expansion. Special consideration is required for drop heights over 80′ because it may greatly increase impact damage to corn, soybeans, and other fragile grains. Consider horizontal conveyors or installing grain decelerators in the down spouts when bucket elevator discharge height exceeds 80′-90′.

Required discharge height

Select bucket elevator height with care. Several heights may be quoted for the same elevator, Fig 2-27. **Effective elevating height** determines power requirements. **Gravity spouting discharge height** is the distance from the elevator bottom to distributor discharge, valves, or perhaps the grain cleaner (e.g. input elevation to gravity spouts). If the elevator bottom is not at grade level (i.e. in a pit or above grade), account for this when determining the required discharge height.

Controls for bucket elevators

If a bucket elevator is overloaded, it can plug and stall. If it stalls and the motor is not stopped, the belt drive pulley can burn through the belt, causing the belt and buckets to fall. Not only is there potential for fire and mechanical damage, but unplugging and repairing an elevator with a broken belt is difficult. Use a motion detection sensor to protect against belt stopping while the motor still operates.

A load indicator or control to prevent overloading the bucket elevator is desirable. An ammeter on the elevator drive motor can indicate load, set off an alarm, or control a relay to stop the feed conveyor. A bin level sensor can activate an alarm and/or stop the elevator motor when the bin is full to prevent grain from backing up the spouts.

Bucket elevators can be equipped with an anti-reverse mechanism. This device prevents the belt from running backwards if the power or drive train fails when the buckets are loaded. Without this mechanism, buckets can be stripped from the belt as they fall and jam in the boot.

Safety

Keep the shields on all bucket elevator drive assemblies. Put guard rails on all service platforms and cages on ladders. Install resting platforms with rails on ladders per OSHA. Have the guy wires positioned, installed, and tensioned by professionals. Elevators supported by towers with stairs are safer and more convenient than elevators and ladders supported by guy wires, but cost more. Climbing to the top of a bucket elevator for servicing, maintenance, or curiosity is dangerous. Forbid climbing unless necessary.

Pneumatic Conveyors

On-farm pneumatic conveyors are usually low capacity conveyors powered by electric motors. High capacity, portable pneumatic conveyors powered by internal combustion engines are usually used for commercial elevator applications, Fig 2-28.

There are two types of pneumatic grain conveyors—positive pressure and negative pressure. Some units combine both types, e.g. grain can be vacuumed from a bin and pushed into a truck, Fig 2-28.

Positive pressure pneumatic conveyors have a positive displacement blower to push air and grain through the pipe, Fig 2-29. Grain flows into the air stream through an air lock.

Table 2-10. Bucket elevator capacities and matching auger applications.

Capacity bu/hr	Matching augers	Applications	Comments
Less than 1,000	4″ augers in dry grain 6″ augers in wet grain 6″ gravity spouts	Small farm needs Well suited to feed processing May serve as wet or dry grain elevator for continuous flow dryer Seed handling	Capacity is too low for grain receiving or load-out.
1,000-2,000	6″ augers in dry grain 8″ augers in wet grain 6″ gravity spouts	Small to medium farms, cash-grain or livestock Can be the primary bucket elevator in a small continuous flow or batch drying system	Lowest recommended capacity for grain receiving and loadout on small livestock or cash-grain farms.
2,000-3,500	8″-9″ augers in dry grain 10″-12″ augers in wet grain 8″ gravity spouts	Medium to large farms. Wet grain receiving and grain loadout. Load-unload for large batch dryers. Primary conveying method for transferring grain in the system.	Recommended for medium to large volumes where a high degree of convenience, reliability, and performance are desired.
3,500-6,000	10″-12″ augers in dry grain 12″-14″ augers in wet grain 10″ gravity spouts	Large farms. Primary grain receiving and loadout conveyor. Load and unload large batch dryer and transfer grain in the system.	Recommended for large to very large farms with multiple handling needs for drying, marketing, and feeding and where high convenience, reliability, and performance are desired.

Pneumatic conveyors can plug. Choose a unit that has easy access to clean out any part of the system. Clean the filters to prevent airflow restriction and capacity reduction. Negative pressure conveyors can suck in a lot of fines and foreign material, especially when moving sunflower. These can plug the filter or build up on the blower vanes and stop the blower.

Positive pressure pneumatic conveyors can be developed into a completely interconnected, push-button, grain conveying system. They adapt well to systems that require relatively low capacity conveying over long distances (greater than 100′) to several different points and directions. Positive pressure pneumatic conveyors are used mainly to deliver grain to storage from the dryer or receiving pit. The primary conveying task is to unload storages, but the blower and injector unit has to be moved to the bin for unloading, then back to a dryer or receiving area for refilling the bin.

Grain can be routed to alternate points by diverting grain flow in a "Y" tube or by shifting a flexible tube to alternate lines at a manifold. The tubes can go above or below ground, up or down grade to follow terrain, and around corners to reach out-of-the-way bins.

Pneumatic conveying systems generally have a lower initial cost than bucket elevator systems, but have higher power requirements and lower capacity. Pneumatic conveying of wet grain at flowrates over 1,500 bu/hr (about the wet grain receiving rate of an 8″ inclined auger) usually requires at least 75 hp. In contrast, a 2,500 to 3,000 bu/hr bucket elevator with an 80′ discharge height can handle wet grain with only 12 to 15 hp. If a tractor is always available when needed, the high horsepower needs may not be a limiting factor. If another engine or electric motor must be purchased, then it is a large factor. Capacities over 1,000 to 1,500 bu/hr may not be realistic for electric motors. Pneumatic conveyors with flowrates over 1,000 to 1,500 bu/hr are usually PTO-driven.

A pneumatic conveyor is not as energy efficient as most conveying systems. Energy requirements are 2 to 2.5 kw-hr/100 bu conveyed; a bucket elevator requires 0.5 to 0.75 kw-hr/100 bu.

Pneumatic conveying works well for relatively low flowrates of dry grain, such as from a continuous flow dryer or dryeration bin to storage, or from storage to a loadout bin. Flowrates of 200 to 500 bu/hr are usually adequate, requiring relatively small distribution tubes (4″-5″), and power needs are easily

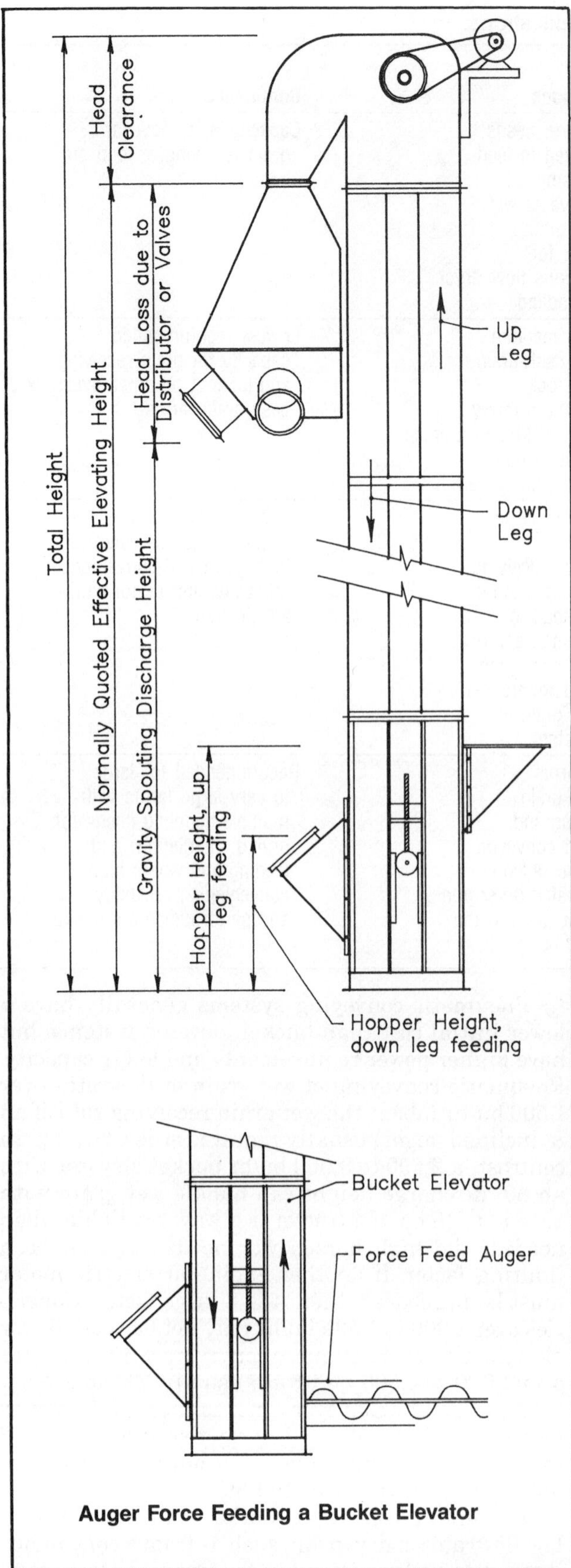

Fig 2-27. Typically quoted bucket elevator dimensions.
When determining required elevator height, consider that the bucket elevator may be in a pit and/or the grade elevation may change from elevator to storage.

Table 2-11. Example pneumatic conveyor capacity.
For a positive pressure pneumatic conveyor with a 4″ diameter line and a 20 hp blower with a ¾ hp airlock. Source: Prairie Agricultural Machinery Institute Test Report No. 1860.

Vertical distance ft	Horizontal distance, ft 25	50	75	100	125	150	175	200
	- - - - - - - -	-	-	bu/hr	-	-	-	-
20	915	876	833	793	755	718	682	649
40	858	821	787	745	709	675	642	610
60	805	772	735	700	667	635	604	574
80	758	726	692	660	628	598	569	541
100	714	684	653	622	593	564	537	510

Table 2-12. Typical pneumatic conveyor capacity and power.
For moving dry grain 150′ horizontally.

Pipe dia.	Capacity bu/hr	Power required hp
3″	400	10
4″	700	15
5″	1,200-1,500	20-30
6″	2,000	40-50

met with electricity. Pneumatic conveyors can unload a batch dryer, but a surge bin is required to quickly unload the dryer.

The conveying rate of a given tube depends on grain type, tube length, number of turns, and change in elevation, Tables 2-11 and 2-12. Capacity may be smaller than expected, because capacity is generally listed for a short horizontal distance and a small lift. Increasing length reduces capacity. Increasing lift reduces capacity considerably and may increase power needs. Select pneumatic conveyor capacity based on required conveyor length and life from present use and future expansion.

One study on grain damage in pneumatic conveyors indicates that damage to corn in a 4″ diameter system is not serious. Damage studies show grain impact can be a problem, but "scuffing" of grain as it slides along a surface is not, as long as pipe sections are properly aligned.

Sharp turns can increase grain damage and wear out the tube rapidly. Make all turns with a large radius. Blowers normally give higher grain velocities than positive displacement pumps, which can lead to higher grain damage. Research indicates that air velocity should not exceed about 5,000 feet per minute (ft/min) to minimize grain damage. Maintain the manufacturer's recommended air-to-grain volume ratio to minimize grain damage. Use a grain decelerator (usually a cyclone) at the discharge, Figs 2-28 and 2-29.

Even though a pneumatic conveyor can easily reach a bin in any location, it is still important to locate bins and equipment in a well planned layout for present use and future expansion.

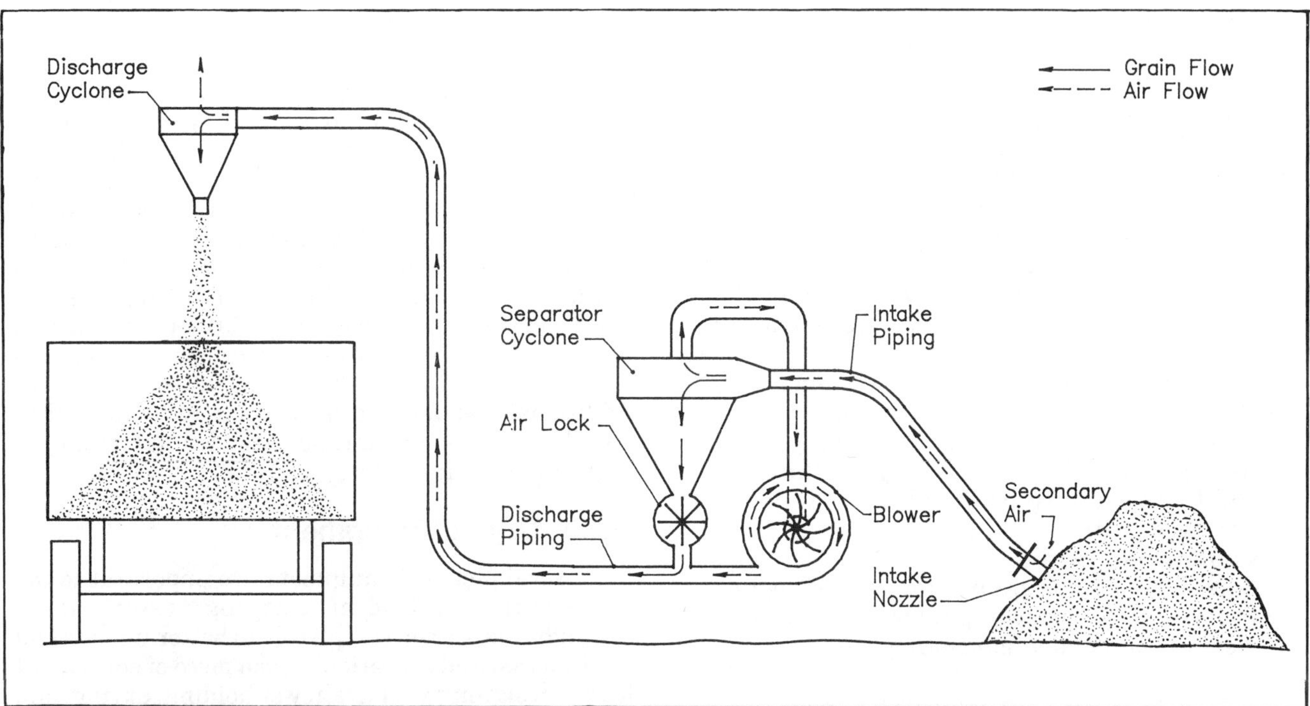

Fig 2-28. Combination positive and negative pressure pneumatic conveyor.
Typical conveyor for commercial elevator application. A blower pulls air and grain through the intake nozzle by negative pressure to the in-stream cyclone where they are separated. Air is pulled through a filter into the blower, compressed, and blown into the grain stream again, pushing grain to the discharge by positive pressure.

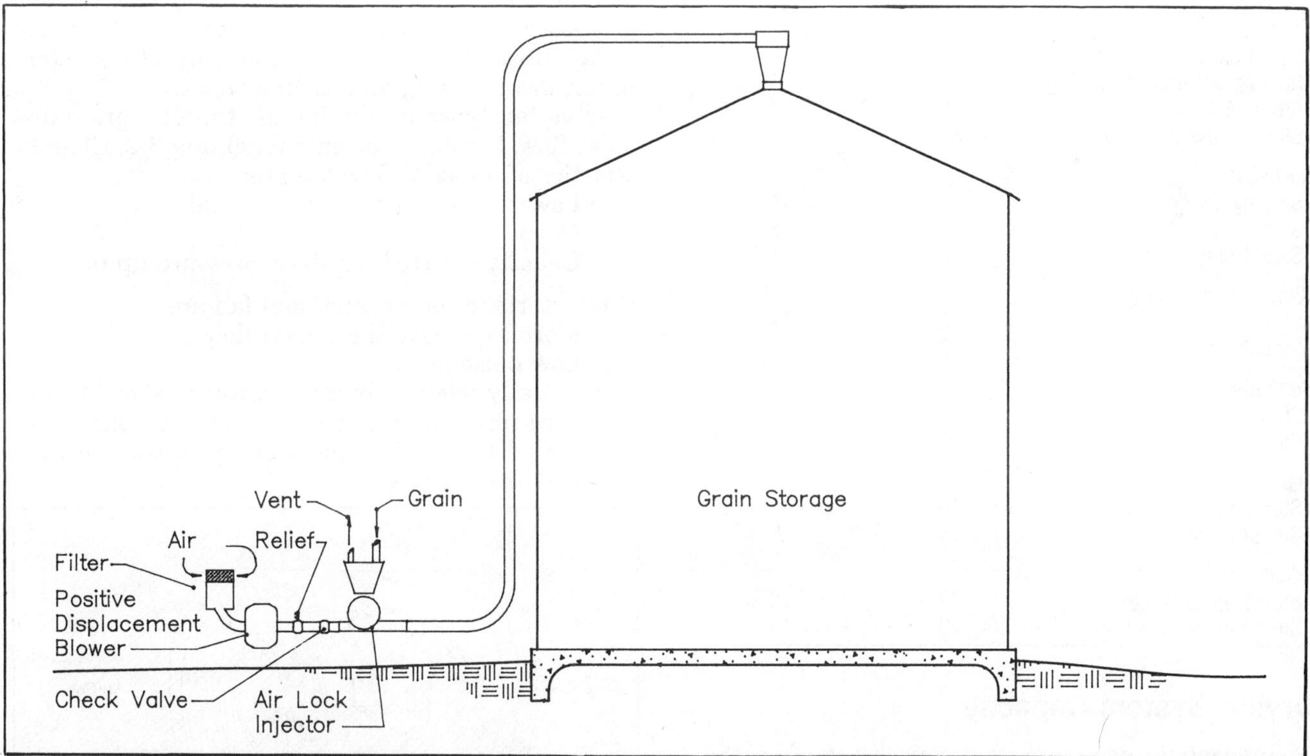

Fig 2-29. Positive pressure pneumatic conveyor.
Typical on-farm pneumatic conveyor. Usually permanently installed and powered by an electric motor.

3. DRYING

Grain harvested too wet for safe storage is usually dried for maximum marketing potential and storage flexibility. Table 3-1 gives maximum harvesting and safe storage moisture content for grains.

A grain drying system is more than just a dryer. It is a group of related processes that receive harvested grain and condition it for marketing or storage. A complete drying system includes wet grain holding, drying wet grain, cooling hot, dried grain, and conveying and handling the grain.

Goals for a well designed drying system are:

- Timely harvest of top quality grain.
- Safe and pleasant working conditions.
- Ability to do important drying jobs efficiently:
  Hold wet grain to be dried.
  Dry wet grain.
  Cool dried grain.
- Capacity to handle grain at the harvesting rate.
- Provisions for easy expansion to future increased volume and harvesting rate.

Table 3-1. Maximum moisture contents for grain harvest and safe storage.
Values for good quality, clean grain and aerated storage. Reduce 1% for poor quality grain.

	Maximum moisture content	
Grain type & storage time	For harvesting %	For safe storage %
Shelled corn and sorghum		
Sold as #2 grain by spring	30	15½
Stored 6-12 mo	30	14
Stored more than 1 yr	30	13
Soybeans		
Sold by spring	18	14
Stored up to 1 yr	18	12
Stored more than 1 yr	18	11
Wheat, oats, barley		
Stored up to 6 mo	20	14
Stored more than 6 mo	20	13
Sunflower		
Stored up to 6 mo	22	10
Stored more than 6 mo	22	8
Flaxseed		
Stored up to 6 mo	15	9
Stored more than 6 mo	15	7
Edible beans		
Stored up to 6 mo	20	16
Stored more than 6 mo	20	14

Drying System Capacity

Inadequate drying system capacity is often the bottleneck in a harvesting-drying-storage system. Provide enough capacity to receive and handle the daily harvest rate. Select equipment that allows for increasing capacity in the future.

Drying capacity depends on the whole drying system, not just the dryer. To determine the average receiving capacity for drying (bu/day), divide the annual grain volume (bu) to be dried by the number of operating days expected in the harvest season. For example, there is a 90% chance that corn can be harvested in 20 to 25 working days out of the 45- to 50-day normal corn belt fall harvest season. To allow for harvesting other crops, tending livestock, and machinery breakdown, 15 corn harvesting days is a safer planning time for many farmers. So 30,000 bu/yr production requires an average receiving capacity of 2,000 bu/day (30,000 bu/yr ÷ 15 days/yr). Additional wet holding facilities may be needed to handle receiving grain from extended harvesting or higher harvesting rates.

Grain Drying Equipment

Several pieces of equipment are important to the successful operation of a drying system. Read through this section completely to better understand the function and operation of each piece of equipment before designing your grain wet holding, drying, and cooling system.

Fans

Selecting a fan is one of the most important considerations for all grain holding, drying, and cooling processes. The fan determines the amount of air that moves through the grain. The amount of air determines grain drying and cooling times.

Two fan types for forcing air through grain are axial flow, Fig 3-1, and centrifugal, Fig 3-2. Characteristics of an axial flow fan are:

- Lower first cost than centrifugal.
- Noisy.
- Usually selected for static pressure up to 3″.

Characteristics of a centrifugal fan are:

- More expensive than axial flow.
- Low noise level.
- Usually selected for static pressure above 4″. For static pressure between 3″ and 4″, compare individual fan performances and pick the fan that best fits your needs.

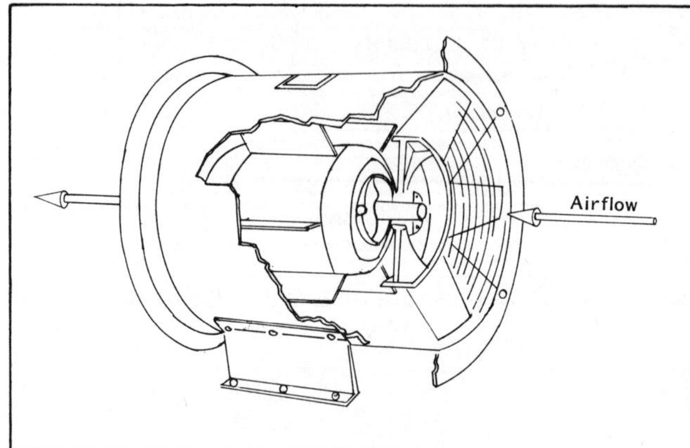

Fig 3-1. Typical axial flow grain drying fan.

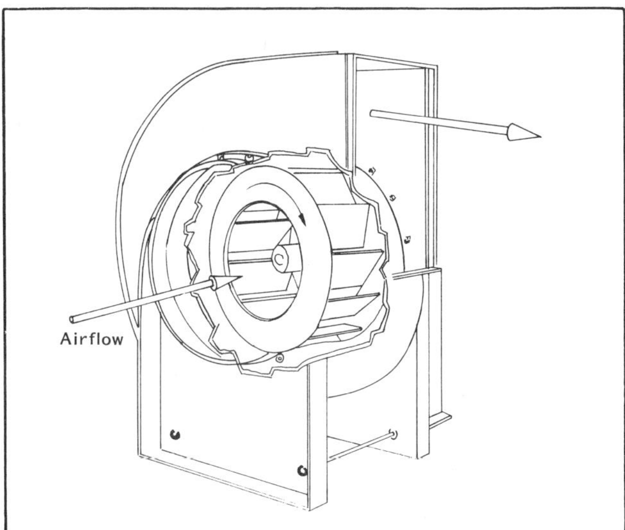

Fig 3-2. Typical centrifugal grain drying fan.

Fan selection

Select a fan primarily for **airflow rate** and **static pressure.** Airflow is measured in cubic feet per minute (cfm) and static pressure in inches of water column (in. of water). Noise level may also be an important selection factor.

The static pressure a fan operates against is affected by:

- Grain depth. Resistance increases linearly with depth. Each added foot of grain adds about the same amount of airflow resistance.
- Airflow through the grain, cubic feet per minute per bushel (cfm/bu). For typical drying and aeration, doubling airflow triples airflow resistance and requires 6 times more fan horsepower.
- Grain type. Airflow resistance depends on grain characteristics (e.g. kernel size and shape) and volume of air moving through the grain. Resistance and power both increase when moving more air through the same grain depth.
- Amount and distribution of foreign material. Foreign material plugs the spaces between grain and increases airflow resistance.
- Grain compaction. The extent of compaction depends on moisture content, filling method, and grain depth.
- Exhaust airflow area out of the bin with positive pressure fans and intake area with negative pressure fans. Provide at least 1 ft^2 of exhaust or inlet area per 1,000 cfm.
- Fan transition and/or duct air velocities.

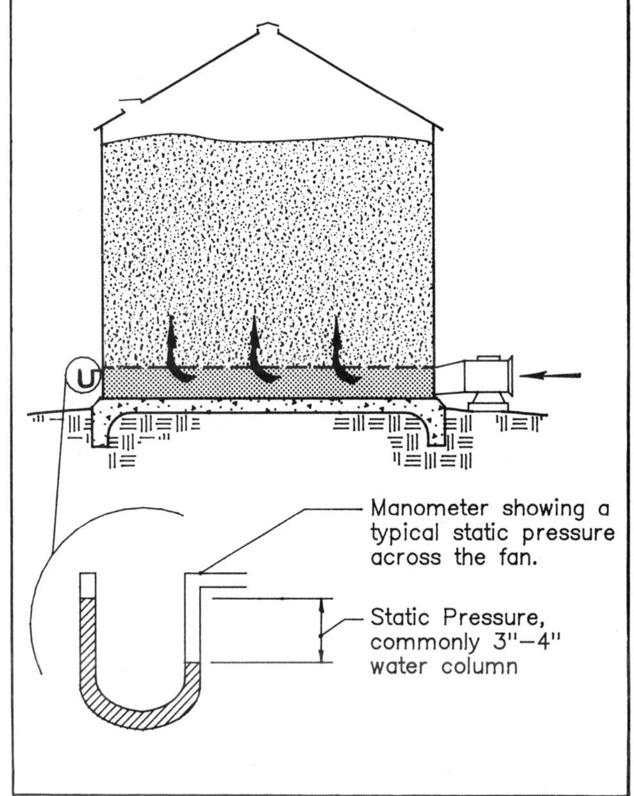

Fig 3-3. Measuring bin static pressures.

Figure 3-3 shows a static pressure measurement often taken in a bin. The manometer (U-tube) at the bottom of the bin shows the static pressure the fan operates against to overcome the resistance to airflow in the system. With a properly designed system, grain is the major source of resistance. The fan selected must operate at the static pressure needed to overcome the airflow resistances in the bin.

Each type and size of fan has unique performance characteristics, presented in a fan curve or fan table as airflow (cfm) vs. static pressure (in. of water). Table 3-2 and Fig 3-4 show example performance characteristics for the **same two fans.** These are just examples and should not be used in planning your system. Use the fan curve or table for your specific fan to evaluate its ability to provide the required airflow. The required fan curve or fan table is available from your fan dealer or manufacturer. Use crop drying fans rated according to AMCA (Air Movement and Control Association) specifications, an independent fan testing laboratory.

Table 3-2. Typical fan tables.

Each fan motor is rated at 10 hp. These tables represent the average performance for 11 axial flow fans and 7 centrifugal fans illustrated in the fan performance section of MWPS-22, *Low Temperature & Solar Grain Drying Handbook*. These tables are only for illustration—use the tables for your fan. Note that axial flow fans generally move more air than centrifugal fans below 3″ and centrifugal fans generally move more air than axial flow fans above 4″. Between 3″ and 4″, they are similar.

	Static pressure (inches of water)								
Fan type	**0**	**1**	**2**	**3**	**4**	**5**	**6**	**7**	**8**
	cfm								
Axial flow	18,000	16,500	15,000	13,500	11,500	7,000	—	—	—
Centrigual	14,700	13,800	13,000	12,100	11,200	10,300	9,100	7,700	5,500

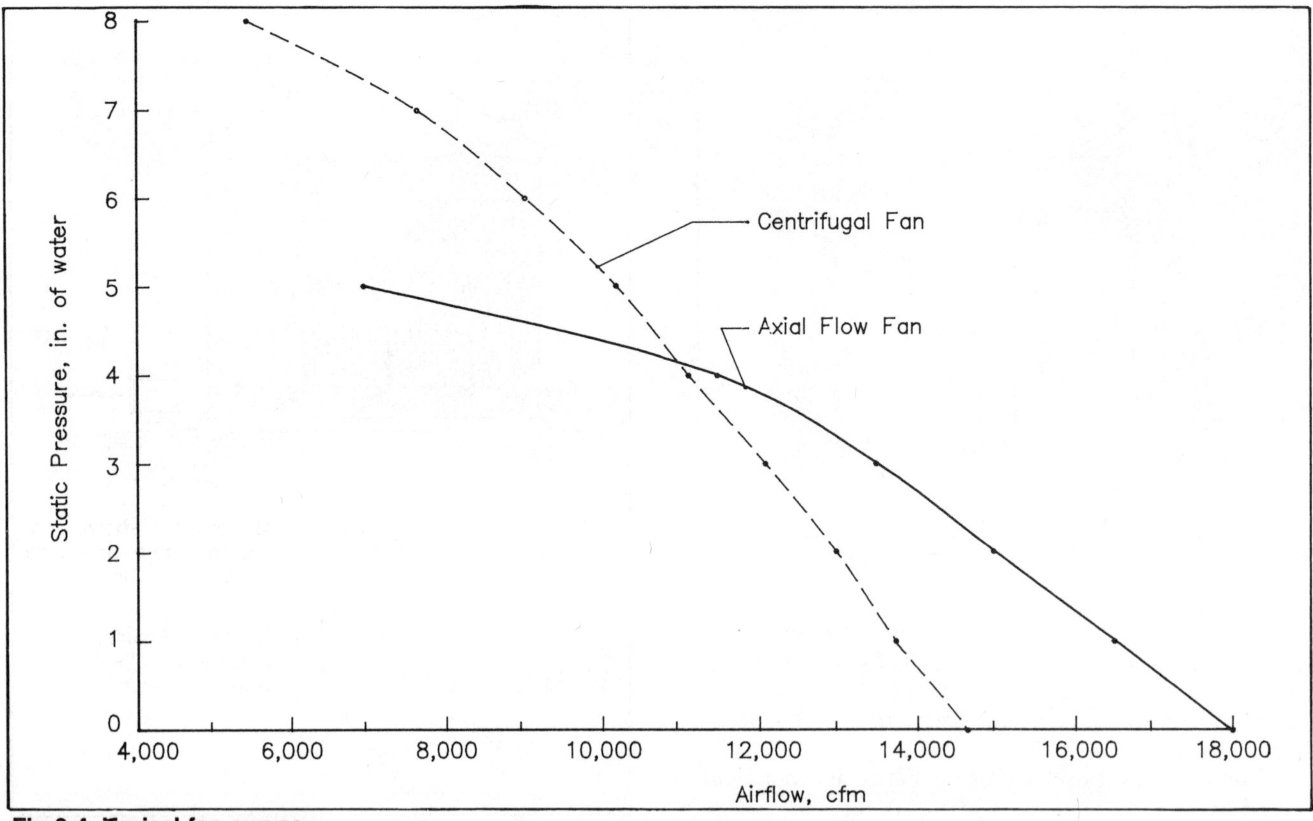

Fig 3-4. Typical fan curves.
Each fan motor is rated at 10 hp. Example fan selection curves based on the average performance of 11 axial flow fans and 7 centrifugal fans. These values are only for illustration—use the curves provided for your fan.

Proper fan selection is based on moving the required amount of air through the grain. Many factors affect the required airflow. For example,:

- Is the fan for drying? If so, is it on a high-temperature or unheated air drying bin?
- Is the fan to be used for dryeration or in-bin cooling?
- Or, is it an aeration fan on a storage bin?

Fans should be selected using either a computer program or from calculations based on specific fan curves or charts. Use these powerful tools to assist you in making major decisions on fan sizing and application. These are available from your extension agricultural engineer or knowledgeable commercial dealer.

Airflow per bushel (cfm/bu) is the most useful characteristic for evaluating a fan because it directly relates to the time required to dry or cool grain. Determine airflow rate per bushel (cfm/bu) by dividing total airflow rate (cfm) by the number of bushels being dried, cooled, or aerated. For a given grain type, grain depth, and bin diameter, total airflow (cfm) can be used to compare different fans and fan arrangements, Table 3-4.

The rated horsepower of a motor is not a good way to compare fans. Fan performance is determined by its design. The horsepower required by a fan is determined by the airflow, static pressure, and fan efficiency. At different combinations of airflow (cfm) and static pressure (in. of water), the horsepower delivered to the fan varies. Shown in Table 3-3 is the calculated fan horsepower for the axial flow fan shown in Table 3-2. This table is shown to illustrate that the required fan horsepower varies considerably with operating conditions and static pressure.

A maximum required horsepower input occurs at a specific combination of static pressure and airflow. The fan motor must be able to operate at the maximum horsepower on a continuous basis. A properly cooled motor can operate at a horsepower above the motor's rating. On the motor nameplate is a service factor which indicates at what level of overload the motor can operate adequately. For example, a motor with a service factor of 1.1 can operate successfully at 10% above its nameplate rating. If the fan motor must operate at a horsepower above the motor's capability, the life of the motor will be reduced. The maximum calculated fan horsepower for the fan in Table 3-3 is 10.6 which is 6% above the rated horsepower. If the service factor for this fan is above 1.06, no problems should occur. Mounting the motor in the airstream improves cooling for maximum motor performance. Problems can occur if the fan motor operates continuously at maximum conditions and is excessively loaded or improperly cooled.

Although fan horsepower is usually not included in fan literature, any reputable fan manufacturer should be able to provide fan horsepower performance characteristics for a fan at different operating conditions.

All motors are rated and have a "motor nameplate" giving the electrical characteristics of the motor. Included on the motor nameplate is the full load current rating of the motor in amps. While a fan is operating

Table 3-3. Calculated fan horsepower.
Fan horsepower depends on airflow, static pressure, and fan efficiency.

	Static pressure (inches of water)					
	0	1	2	3	4	5
cfm	18,000	16,500	15,000	13,500	11,500	7,000
fan hp	8.6	9.5	10.2	10.6	10.6	9.9

an electrician can measure the current to determine the load characteristics of the motor. If this is above the full load current rating, the motor is overloaded.

The effect of corn depth on airflow and static pressure is shown in Table 3-4. As depth increases, static pressure increases, but airflow decreases reducing the drying capacity. Table 3-4 also shows the effect of adding up to 3 fans to a bin.

Table 3-4 is an example that illustrates principles, but does not describe specific equipment or recommendations. Values in Table 3-4 are based on a 30′ diameter drying bin and the 10 hp axial flow and centrifugal fans from Table 3-2.

Consider grain depth when selecting fans because extra depth increases airflow resistance which decreases drying rate. Drying capacity is directly proportional to airflow. For a given drying air temperature, any diameter bin, grain depth, or number of fans, the relative drying capacity can be determined by comparing total airflow. For example, from Table 3-4 one 10 hp axial flow fan at a grain depth of 6′ dries 1.30 times (15,275 cfm ÷ 11,745 cfm) or 30% faster than grain 18′ deep. To compare different grain depths, bin diameters, and fan combinations, you must compare airflow rate (cfm/bu).

For a given bin, grain depth, and air temperature, adding an additional fan increases the airflow and drying rate. Compare airflow rates at a given depth in Table 3-4 to determine the effect of adding more fans. For example, the relative drying capacity of two 10 hp axial flow fans at a grain depth of 6′ is 1.60 times (24,380 cfm ÷ 15,275 cfm) or 60% greater than one fan at a 6′ grain depth. Consider the cost and benefit of increased drying capacity when deciding to add another fan to a drying bin.

Heaters

Drying air temperature also affects drying capacity. Increasing drying air temperature increases drying capacity, e.g. increasing the temperature from 120 F to 180 F almost doubles drying capacity (16.5 ÷ 8.5). Table 3-5 **estimates** drying capacity of air at various temperatures. For example, at 160 F relative drying capacity is 1.62 or 62% more than at 120 F (13.8 ÷ 8.5), so approximately 60% more grain is dried per unit of time with 160 F air rather than 120 F air.

Increasing drying air temperature in a given dryer usually increases fuel efficiency. As fuel efficiency increases, more bushels of grain are dried per gallon of fuel.

Table 3-5. Estimated drying capacity of heated air.
Ambient air at 50 F and 65% relative humidity.

Drying air temperature, F	Drying capacity, corn dried from 24% to 14% m.c. bu/hr per 1,000 cfm	Relative drying capacity
50*	0.64	1.00
120	5.06	8.50
140	6.59	11.1
160	8.18	13.8
180	9.81	16.5

*Natural air at 50 F and 65% relative humidity.

Table 3-6 shows the effect of drying air temperature on drying capacity in bushels per hour for a 30′ bin with one 10 hp centrifugal fan. At a given drying air temperature, increased grain depth decreases drying capacity. Relative drying capacity increases at a given grain depth as drying air temperature increases. For example, at a 6′ grain depth 1.60 times (109 bu/hr ÷ 68 bu/hr) or 60% more grain dries at 160 F than at 120 F.

Table 3-4. Example fan performance at different grain depths.

		Axial flow			Centrifugal		
No. of fans	Corn depth ft	Static pressure in. water	Total airflow cfm	Airflow rate cfm/bu	Static pressure in. water	Total airflow cfm	Airflow rate cfm/bu
One	2	0.7	16,890	14.9	0.6	14,190	12.6
	6	1.8	15,275	4.5	1.6	13,355	3.9
	10	2.7	13,935	2.5	2.4	12,645	2.2
	14	3.4	12,755	1.6	3.1	11,980	1.5
	18	3.9	11,745	1.2	3.7	11,450	1.1
Two	2	1.8	30,555	27.0	1.5	26,845	23.7
	6	3.7	24,380	7.2	3.5	23,370	6.9
	10	4.5	19,350	3.4	4.9	20,755	3.7
	14	4.8	16,355	2.1	5.9	18,440	2.3
	18	5.0	14,180	1.4	6.5	16,770	1.7
Three	2	2.9	41,000	36.3	2.5	37,705	33.3
	6	4.6	27,515	8.1	5.3	29,975	8.8
	10	5.0	21,020	3.7	6.5	25,260	4.5
	14	5.2*	16,935*	2.1*	7.3	21,605	2.7
	18	5.3*	15,010*	1.5*	7.7	18,655	1.8

*These static pressures exceed the maximum value in the fan tables and curves. Therefore, the total airflow and airflow rate values shown are estimates.

Table 3-6. Effect of drying air temperature on drying capacity.
Bin diameter is 30′. One 10 hp centrifugal drying fan from Table 3-2 is used.

Corn depth ft	Drying capacity, bu/hr				
	50 F[a]	120 F	140 F	160 F	180 F
2	9.1	72	94	117	141
6	8.5	68	83	109	131
10	8.1[b]	64	84	104	125
14	7.6[b]	61	79	98	117
18	7.3[b]	58	76	94	113

[a]Natural or unheated air at 50 F and 65% R.H.
[b]Required drying time for 24% m.c. corn exceeds safe natural air drying times. The grain may spoil. See MWPS-22, *Low Temperature & Solar Grain Drying Handbook*.

LP gas (propane) is a common fuel for high temperature bin dryers. To provide the necessary heat output a vaporizer is usually required, especially in cold weather. Most burners have self-contained vaporizers. As fuel costs increase, biomass burners may become more common. Leave space around a new high temperature bin dryer for fuel storage, access for refueling, and possibly a biomass burner.

In this book, biomass refers to crop residues and wood. Crop residues (corn stover and cobs, straw, etc.) are renewable and can provide a reliable, low-cost energy source. Biomass furnaces have the greatest potential with high temperature dryers. Use for grain drying depends on collecting, transporting, and storing the crop residue; cost; and the time to operate the system during rapid harvest.

With continued development, crop residue furnaces and collection equipment have tremendous potential for high temperature drying. In a new grain system, include space in the layout for a furnace and crop residue handling and storage.

As energy costs increase, there is interest in alternative fuels for grain drying. Solar energy and crop residues have been used and will likely be used more in the future.

Solar energy is adaptable to low temperature dryers, which need low heat input over extended periods. These drying methods tolerate intermittent or variable heat input. Solar collectors can be added to existing or planned farm structures. Simple collectors are efficient at the low temperature rises needed for low temperature drying.

Low temperature drying is weather dependent, and in some areas and some fall seasons, supplemental heat may be desirable. Additional heat may be required to achieve the desired final moisture content. If a solar collector is used for more than one purpose, such as grain drying and shop heating, it helps offset the collector cost. Refer to MWPS-22, *Low Temperature & Solar Grain Drying Handbook*, for more discussion.

Conveyors

A well organized grain conveying system is needed to move grain between wet grain holding, the dryer, and grain cooling. Conveyor speed is determined by the process demands. Fast, high capacity conveyors are typically used with batch drying systems. Slower, low capacity conveyors are needed for continuous flow drying systems. See Chapter 2 for more about conveyors.

Condensation Problems

If the season's last batch is dried and stored in a high temperature bin dryer, a perforated wall liner or perforated tubes spaced 9″ o.c. around the inner wall are needed to minimize sidewall moisture condensation. Perforated tubes or a wall liner should extend about ¾ of the way up the wall height. Install perforated tube ends to start below the perforated floor and the wall liner no more than 1′ above the perforated floor. If the last batch of the season is dried with low temperature air, properly cooled, and/or temporarily stored for marketing or feeding before spring, this protection is probably not needed.

Condensation can occur in conveying equipment when moving hot grain from a high temperature dryer, particularly in colder weather. For example, condensation in a bucket elevator can accumulate in the boot and freeze. Check equipment frequently for condensation and wet trash.

Condensation also tends to form in downspouts and distributors and run into bins. Consider installing a manual-, gravity-, or spring-operated flapper valve at the bin end of downspouts to keep air from flowing up into them. Or, install exhaust fans in the bin roof to keep the bin headspace under a slight negative air pressure to minimize air from moving up the downspouts. Size these fans for 25%-50% more air than the cooling fan rated at a static pressure of at least ⅛″.

Grain Cleaners

Cleaning grain before drying improves airflow and drying capacity. Fines, foreign material, and broken kernels can collect on a perforated floor or in the space between kernels blocking airflow and reducing drying capacity. See Chapter 2 for more about grain cleaners.

Moisture Testing

Moisture testing freshly dried, hot grain requires special procedures because warm grain can cause the tester to give false readings. This can be corrected with an adjustment factor. Determine the adjustment factor for your tester by quickly cooling the grain to room temperature and take a moisture reading, seal the rapidly cooled sample in a mason jar, and read again in 24 to 48 hr. The difference between this reading and the original reading is the adjustment factor for your meter. This factor is a constant for a particular moisture meter. Use this adjustment factor to correct future readings.

Holding Wet Grain

With extended favorable weather, dryer capacity can limit harvest rate. But, extra capacity wet grain holding facilities can extend drying to days when it is unfavorable for harvesting.

Effective ways to hold wet grain before drying range from a few hours in wagons or trucks to a month or more in the top of a low temperature drying bin. Between these two extremes, wet grain is commonly held for several days in bins with fans to keep grain cool, Fig 3-5.

With all high temperature dryers, wet grain holding accumulates wet grain ahead of the dryer, so the dryer can operate automatically and continuously for up to 24 hr/day. This is usually one hopper-bottom bin referred to as the operating wet holding bin. Additional or extra wet grain holding, usually in separate bins, allows longer harvesting days, allows drying the rapid harvest of a custom operator, and lets harvest continue if the dryer stops. Several days of drying capacity can accumulate in well aerated wet holding facilities for drying during poor harvesting weather or while waiting for the custom operator to return. Well planned wet grain holding facilities greatly increase the flexibility and effectiveness of a harvest-to-storage system. Size wet holding capacity to use the dryer's maximum capacity and handle grain at the maximum harvesting rate.

Example 3-1:

Determine the required minimum operating wet grain holding bin and drying capacity to handle a harvesting rate of 400 bu/hr. Assume harvesting is conducted in 16 hr each day and all harvested grain is dried in the same day.

Solution:

Daily harvesting capacity is:

400 bu/hr × 16 hr = 6,400 bu

Assume the dryer runs 24 hr/day, the minimum drying capacity is:

6,400 bu ÷ 24 hr = 267 bu/hr

Minimum wet holding capacity is the difference between the volume of grain harvested and dried during the harvesting period each day. Drying capacity during the 16 hr of harvesting is:

267 bu/hr × 16 hr = 4,272 bu

Minimum wet holding capacity is:

6,400 bu − 4,272 bu = 2,128 bu

Plan for at least 50% more drying or wet holding capacity to handle longer harvesting times, carry-over from previous day, reduced drying capacity, or down time for repairs. For this example, select an operating wet grain holding bin capacity of 3,200 bu (2,128 × 1.5).

This allows the dryer to keep up with harvesting and not create a bottleneck. All grain harvested each day will be dried in the same day. The wet holding bin will be empty and ready for the next day's harvest.

Grain respiration and fungi and bacteria activity cause wet grain to start heating almost as soon as it is harvested. The heating rate depends on grain moisture content, temperature, and physical damage.

Aerate wet grain to prevent rapid temperature increases that can virtually destroy the grain. Aerate wet grain to be held longer than 12 hr to keep it cool. Aerated bins require properly sized fans and good air distribution. Provide wet holding bins with at least ½ cfm/bu (cubic feet of air per minute per bushel) to cool the full grain depth. For example, a 2,000 bu wet grain holding bin needs a fan that can deliver at least 1,000 cfm (2,000 bu × ½ cfm/bu).

Although perforated floors distribute air best, they are expensive and not feasible in hopper-bottom bins. Properly designed duct systems work well in wet holding bins. Extension agricultural engineers or knowledgeable commercial representatives can help you select the correct fan and air distribution system for wet grain holding bins.

Wet holding time before drying can affect the grain's overall storage life. Table 3-7 relates moisture content and temperature to overall storage life for shelled corn. Table 3-7 is also valid for grain sorghum. There is limited storability data for other grains, but the same general principles apply.

Completely empty unaerated wet holding bins every day to reduce heating and deterioration. Empty aerated wet holding bins so that no more than ½ of the maximum storage life is used up in wet holding. See Table 3-7. For example, according to Table 3-7, the maximum storage time for 24% m.c. corn at 55 F grain temperature is 16 days. The allowable holding time in an aerated wet holding bin is half of the maximum storage time, which is 8 days (16 ÷ 2). Therefore, a holding bin with 24% m.c. corn should be emptied at least every 8 days.

Table 3-7. Maximum storage life for shelled corn.

Values for corn held at a constant temperature, i.e. with respiration heat removed by aeration. These values are **not** the time allowed for wet grain holding—they **are** the total acceptable storage life from corn harvest to the time it is used. Do not use all the storage time on your farm—save some storage time for the next users so they have a quality product to work with. For example, 26% MC (moisture content) corn held at 60 F for 4 days used ½ of its life (4 days/8 days). If then dried to 15½% MC and held at 50 F, the corn has only 225 days remaining (½ × 450).

The values are based on ½% dry matter loss by weight—a one grade decrease or ½% of the corn dry matter is lost to deterioration. Such grain is usually visibly moldy.

w.b. = wet basis.

Developed from Thompson, *Transactions of ASAE 333-337*, 1972.

Grain temperature F	Moisture content, % w.b. 15½	18	20	22	24	26	28	30
	days							
30	2,276	648	321	190	127	94	74	61
35	1,517	432	214	126	85	62	49	40
40	1,012	288	142	84	56	41	32	27
45	674	192	95	56	37	27	21	18
50	450	128	63	37	25	18	14	12
55	299	85	42	25	16	12	9	8
60	197	56	28	17	11	8	7	5
65	148	42	21	13	8	6	5	4
70	109	31	16	9	6	5	4	3
75	81	23	12	7	5	4	3	2
80	60	17	9	5	4	3	2	2

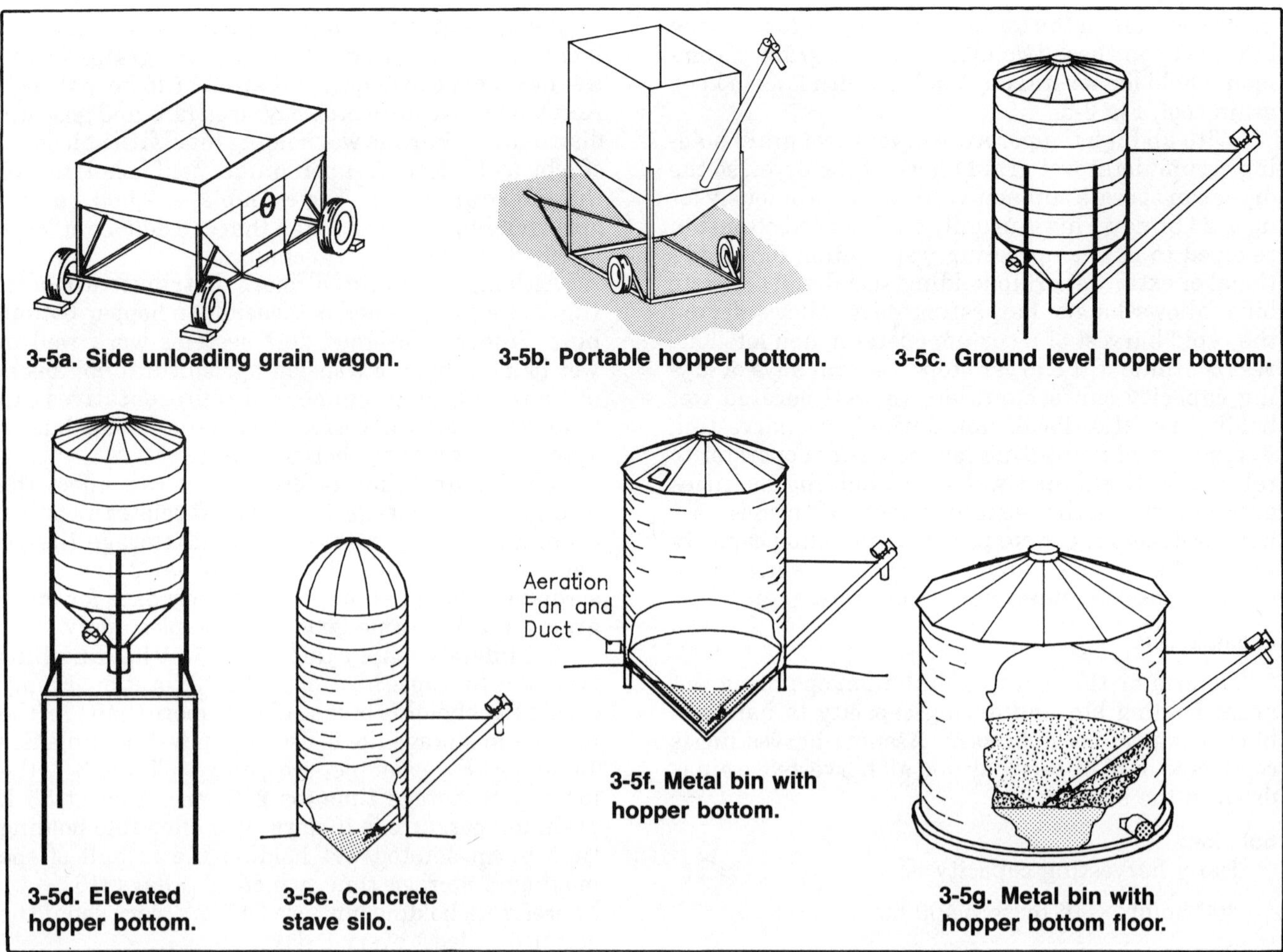

3-5a. Side unloading grain wagon.
3-5b. Portable hopper bottom.
3-5c. Ground level hopper bottom.
3-5d. Elevated hopper bottom.
3-5e. Concrete stave silo.
3-5f. Metal bin with hopper bottom.
3-5g. Metal bin with hopper bottom floor.

Fig 3-5. Wet grain holding.

Grain Cooling

Grain dried in any heated air dryer must be cooled. Grain can be rapidly cooled immediately after it is dried or delayed cooling methods can be used to reduce fuel costs, increase dryer capacity, and reduce stress cracks. The cooling method can affect the type, operation, and management of the dryer system. Consider the cooling method when selecting a dryer.

With delayed cooling, hot grain is usually transferred from a high temperature dryer to a separate bin to be cooled. Additional handling equipment and cooling bins with a fan and perforated floor or ducts are usually needed for delayed cooling. In-bin cooling, dryeration, and combination high/low temperature drying are the 3 delayed cooling methods. Delayed cooling effectiveness increases as drying air temperature increases. The minimum drying air temperature to consider delayed cooling is 120 F.

In-Bin Cooling

In-bin (or in-storage) cooling is the simplest delayed cooling method and is suitable with any type of high temperature dryer. Compared with rapid in-dryer cooling, in-bin cooling can reduce fuel costs at least 10% and increase dryer capacity about 33%. For a comparison of relative operating costs, see Table 3-11.

The general operating procedure is to stop high temperature drying about 1% above the desired final moisture content. Transfer grain to storage and cool continuously to remove some additional moisture. Because cooling is not delayed very long, cooling removes about 0.1 to 0.15 percentage points for each 10 F reduction in grain temperature (about half as much as dryeration). After cooling, grain is usually stored in the bin.

Start in-bin cooling as soon as grain is in the cooling bin. If cooling is delayed, especially in cold weather, moisture condenses on the sidewalls and accumulates in the grain, which can cause spoilage. In-bin cooling can be delayed if outside temperatures are above 50 F to allow grain to steep and increase the amount of moisture removed.

Size the fan so cooling is at least as fast as drying. For example, for a 200 bu/hr drying capacity, size fans to cool at least 200 bu/hr when the bin is full. Determine the **minimum** airflow (cfm) by multiplying the desired cooling rate by 12 cfm/bu/hr. For a 200 bu/hr dryer, the desired cooling rate is 200 bu/hr and the minimum fan size is 2,400 cfm (12 cfm/bu/hr × 200 bu/hr).

The fan must deliver at least 2,400 cfm at the maximum grain depth. Select fans for in-bin cooling with generous capacity. You will need to determine

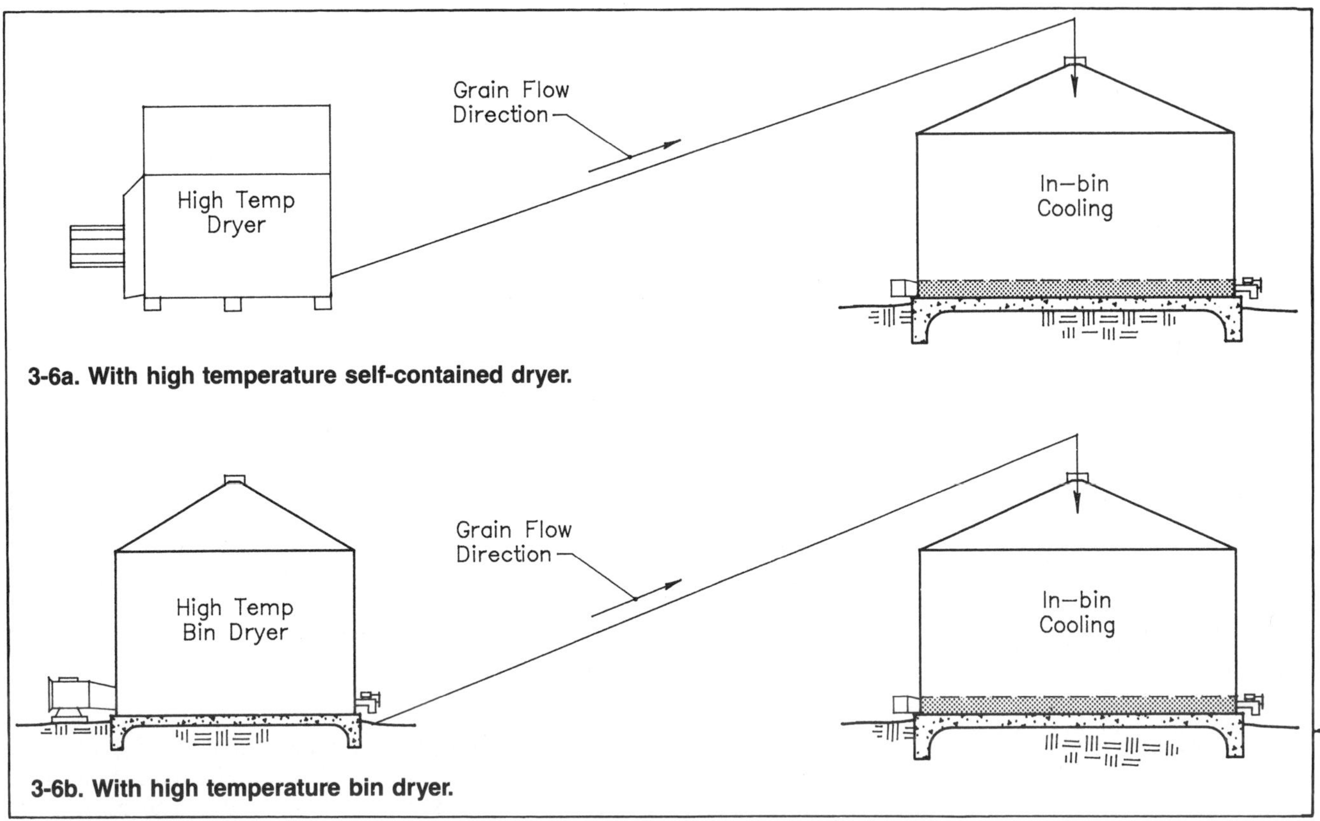

Fig 3-6. In-bin cooling.
Grain dried in a high temperature dryer or bin and cooled in a separate storage.

the static pressure to select a fan. State extension agricultural engineers and knowledgeable grain drying equipment dealers can estimate static pressure for specific grain cooling bins.

An aerated storage bin can be used for in-bin cooling if filled gradually. With 1/10 cfm/bu aeration capacity when full, fill the bin over 6 or more days. With 1/5 cfm/bu aeration capacity, fill over 3 or more days.

Bins with fully perforated floors provide the best air distribution for in-bin cooling. Partly perforated floors or ducts are adaptable, but air movement through the grain may be less uniform. With partial floors or ducts, it is safer to partially cool the first 2′-3′ of grain in the dryer.

Dryeration

Dryeration is an energy efficient method of delayed cooling and drying completion. Compared with immediate, rapid cooling in the dryer, dryeration reduces fuel costs 15%-30%, increases drying capacity 50%-70%, and results in fewer stress cracks, less brittleness, and improved color (bloom) and millability. See Table 3-11.

With dryeration, hot grain is moved to the cooling bin immediately after drying. But, cooling is delayed at least 4 hr and preferably 12 hr for steeping or tempering while the moisture content equalizes in the kernel. This results in less physical stress during cooling and more moisture removal. After cooling grain in the dryeration bin, transfer it to storage.

An important dryeration concept is all hot grain delivered to a dryeration bin must temper at least 4 hr before it is cooled. Two possible filling and cooling methods that assure that all grain steeps the required time are:

- Fill the dryeration bin with one day's dryer operation (up to 24 hr) and delay starting the cooling fan until after the last hot grain is delivered. The bottom grain must temper for at least 4 hr before starting the cooling fan. If the bin cools in 10 to 12 hr, the top grain tempers 10 to 12 hr before the cooling front reaches it.
- Begin filling the dryeration bin and delay starting the cooling fan at least 4 hr and usually 6-12 hr after the first hot grain enters the bin. Delaying the start of the cooling fan more than 4 hr allows the moisture in the kernel to equalize more, resulting in less physical stress and possibly more moisture removal. Cycle the fan on and off with a percentage timer, so the cooling front only reaches grain that has tempered at least 4 hr.

With dryeration, air is moved either up or down through the bin to cool grain. As the grain cools, the air warms to the grain temperature. As the air warms, it picks up moisture from the grain. This warming and drying happens over a short distance in the grain, forming a cooling front that gradually moves through the bin. Air moving through warm grain ahead of the cooling front does not significantly change the warm grain's temperature or moisture

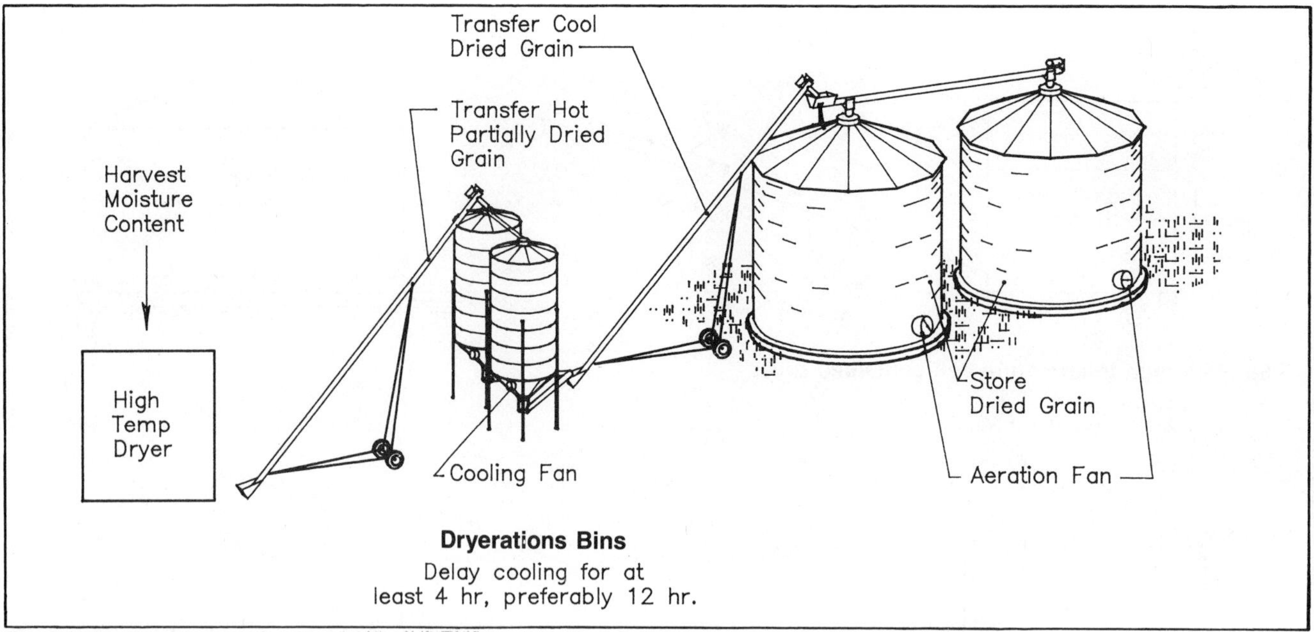

Fig 3-7. Example two bin dryeration system.
With proper grain steeping and cooling in the dryeration bins, 2 to 3 percentage points of moisture content should be removed. See Fig 3-8 for the typical operating procedure for a two bin system.

content, so warm grain still tempers even though air is moving through it. If air moves upward, the cooling fan can run while hot grain is being added, Fig 3-8. However, the last grain into the bin must steep at least 4 hr before being cooled.

The cooling and drying effect of dryeration reduces grain moisture content about 0.2 to 0.25 percentage points for each 10 F reduction in grain temperature. So, the high temperature dryer can be stopped when grain moisture content is 2 to 3 percentage points higher than the desired final moisture content.

Dryeration works best with two dryeration bins and additional storage bins. The most convenient cooling bin cycle is usually 48 hr for loading, steeping, cooling, and unloading. Add hot grain from the dryer to the first bin for up to 24 hr, then divert loading to the second bin. Start the cooling fan after the first grain delivered to the bin has tempered, usually 6 to 12 hr. Finish cooling and unload the first bin during the next 24 hr and begin the cycle again, Fig 3-8. With two dryeration bins, filling and unloading alternates between bins. While grain in one bin finishes cooling and is unloaded, hot grain is loaded into the other bin.

As shown in Fig 3-8, the fill level and the cooling front both advance upward in the bin. If the cooling front overtakes the filling level, the last grain loaded into the bin will not properly steep before cooling. The rate at which the cooling front moves depends on the airflow rate. If the cooling front moves faster than the filling level, reduce the airflow rate with intermittent fan operation. A percentage timer, time clock, or other electrical control device can operate the fan intermittently.

For dryeration, provide airflow of at least ½ cfm/bu and up to 1 cfm/bu when the bin is full. The higher the airflow, the faster it cools: ½ cfm/bu cools hot grain in about 24 hr—1 cfm/bu in about 12 hr. For example, for a 3,000 bu cooling bin, the minimum required airflow is 1,500 cfm (3,000 × ½ cfm/bu).

You can also base required airflow on dryer capacity. Steeped hot grain is usually cooled at the same rate that the dryer fills the bin, so cooling rate and airflow (cfm) depend on dryer capacity. A good rule of thumb is to multiply the maximum dryer capacity, bu/hr, by 12 to get minimum airflow, cfm. For example, if the maximum dryer capacity is 400 bu/hr, required minimum airflow is 4,800 cfm (400 bu/hr × 12 cfm/bu/hr).

While tempering and cooling grain with dryeration, condensation usually accumulates on and in grain near the bin walls, particularly during cold weather. Tempering for more than 16 hr in below freezing weather can lead to frozen condensation. Condensation will cause problems if grain is stored in the dryeration bin. But, when grain is transferred from dryeration to storage, the process of unloading and conveying mixes small quantities of wet grain with dry grain, so it stores well. Dryeration requires a convenient, well planned grain transfer system. Get design help from your state extension agricultural engineer, a consulting engineer, or commercial representative.

Commercial hopper-bottom bins with fans and ducts for cooling, and flat-bottom bins with fans and perforated floors are both used for dryeration. With flat-bottom bins, a cone of grain can be left in the bin between batches to form a hopper. Check the cone regularly for surface fines that might spoil or block airflow. If grain will be stored in the bin, remove the cone with a sweep auger before the season's last filling to remove any accumulated moisture and fines.

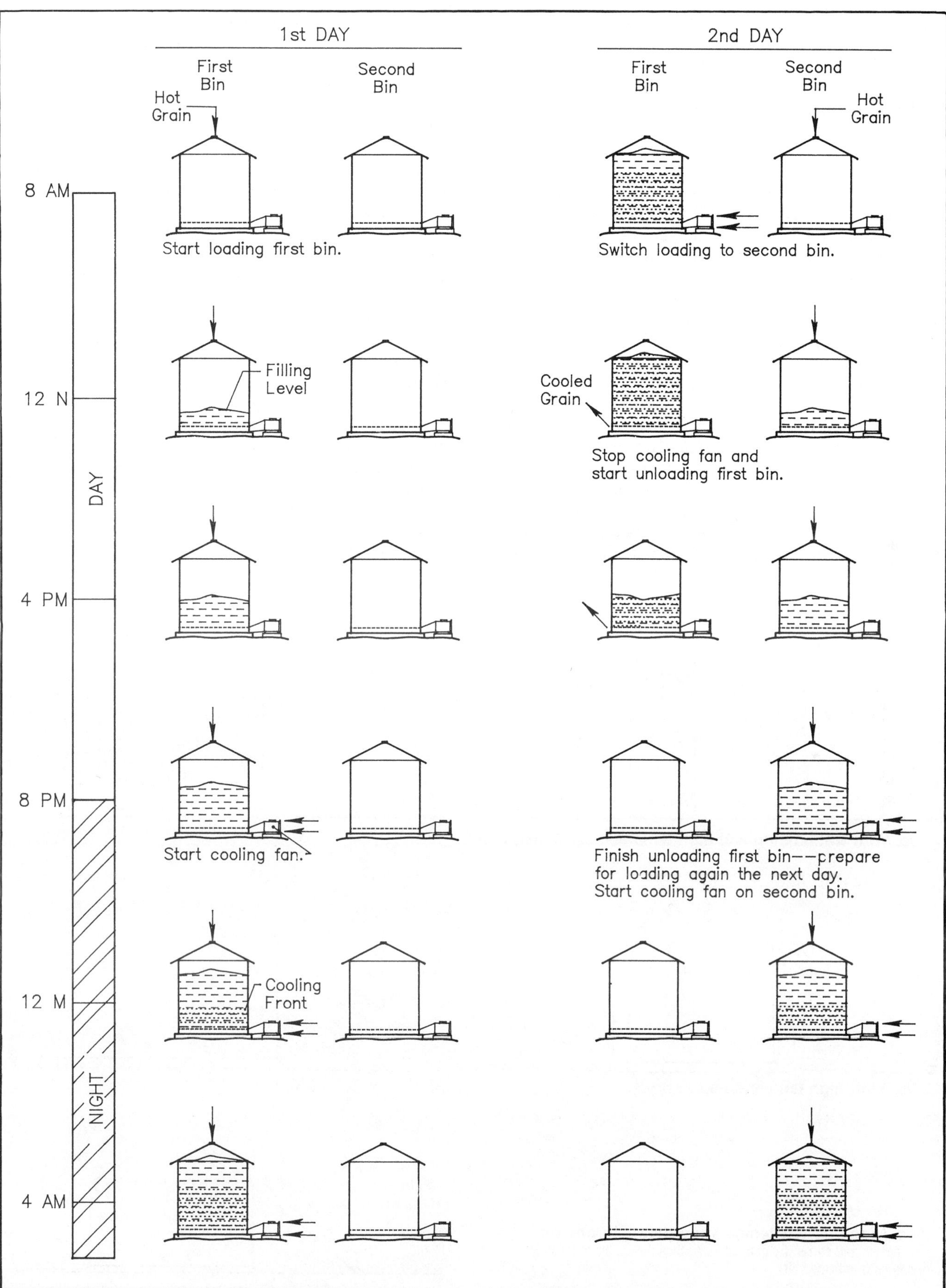

Fig 3-8. Typical dryeration 24 hr cycle with 2 bins.

Size cooling bins to store at least the maximum dryer output for the cooling cycle—usually 16 to 24 hr. Remember that dryer capacity increases 50%-70% with dryeration. Consider oversizing cooling bins up to 50% (1.5 times daily drying capacity) to provide flexibility if drying capacity increases. For example, a high temperature dryer dries up to 2,500 bu in a 16 hr/day drying period. Provide two cooling bins with a capacity of at least 3,750 bu each (2,500 bu × 1.5). The extra capacity provides flexibility to change the drying operation. If drying is extended to 24 hr/day in the previous example, then 3,750 bu of capacity is needed. This can be handled by the planned cooling bins.

Do not let cost or complexity discourage you from considering dryeration with high temperature drying. Even though more equipment and management are required, the extra energy savings, dryer capacity, and grain quality can more than offset the disadvantages. Improved grain quality, especially in corn, is frequently overlooked because the market does not usually pay any more for dryeration-cooled grain at this time. But, if you have a high temperature dryer and have **any storage, handling, or marketing problems with fines and broken grain,** seriously consider dryeration. Remember, dryeration increases drying capacity, decreases drying costs, and maintains grain quality.

Combination High Temperature/Low Temperature Drying

Combination high temperature/low temperature drying is discussed here as a cooling method because grain cooling is separate from the high temperature dryer.

Combination drying is most adaptable for crops harvested too wet for safe low temperature bin drying. It is usually not used for wheat or soybeans but works well for corn. Wet corn (above 22% moisture content) can be dried with heated air to a moisture content of 22% or less. The partially dried corn is delivered hot to a low temperature drying bin. Start the drying fans immediately. This will cool and dry the corn to about 21% moisture content, where low temperature drying is more reliable. Grain is left in the bin for storage.

Equipment needed includes any high temperature bin or self-contained dryer, low temperature drying bins, and a well organized conveying system. A grain cleaner is recommended for better airflow and more reliable low temperature drying.

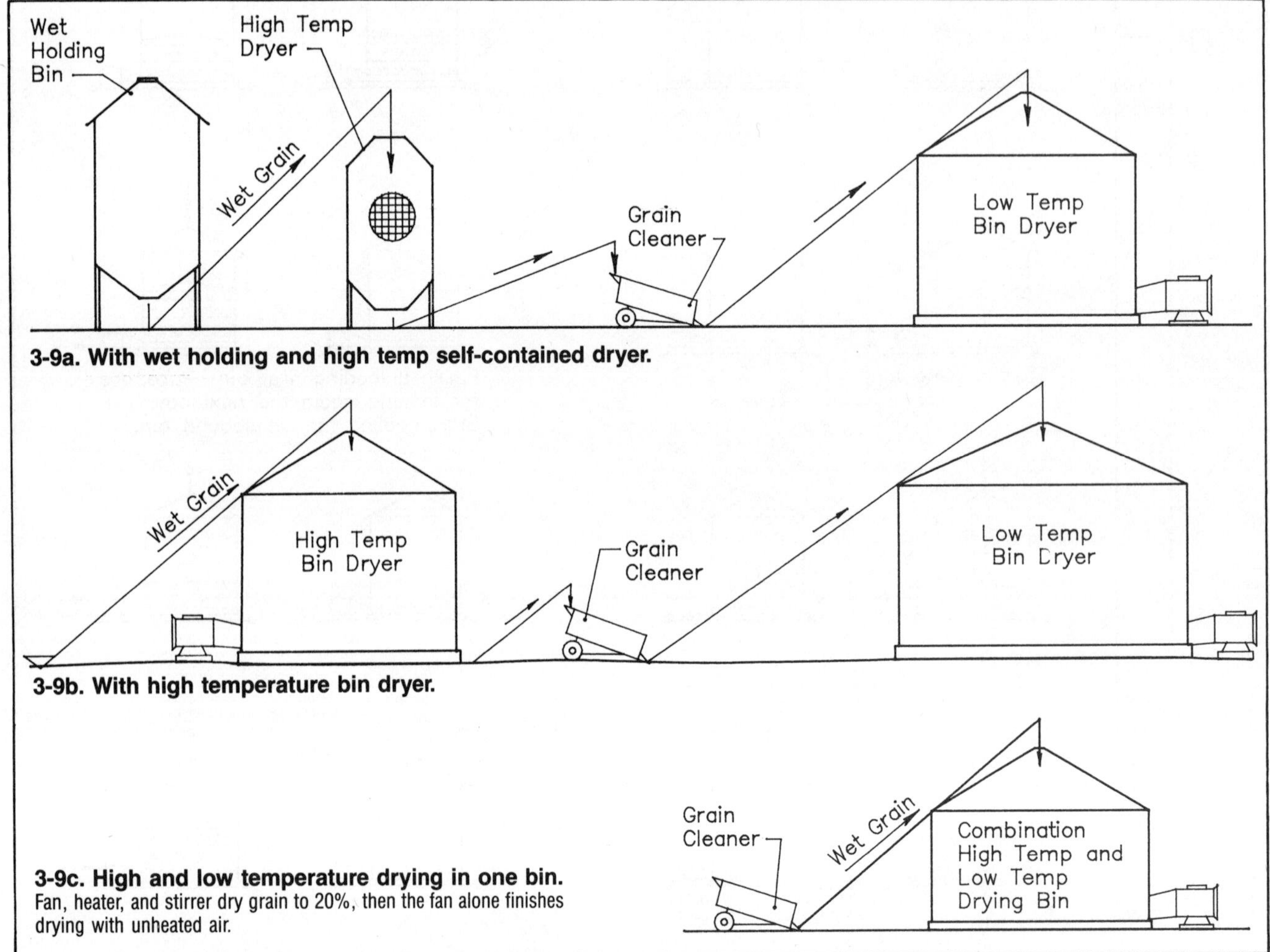

Fig 3-9. Combination drying.
Arrows indicate grain flow direction.

Grain conveying requires thorough planning, Figs 3-9a and 3-9b. Above 22% moisture, grain goes first to the high temperature dryer. Below 22%, it goes directly to the low temperature dryer.

A combination system usually costs more than a single drying method, because all storage bins are equipped for low temperature drying and two drying methods and flexible grain handling are needed.

A bin-batch dryer with stirring equipment can be used for combination drying for the last bin full of the season, Fig 3-9c. Grain with a high moisture content can be dried with heated air to a moisture content that can be safely dried with unheated air. See the drying systems section for more about unheated and heated air dryers.

Combination drying gives both the high drying capacity and flexibility of high temperature drying and the energy savings and superior grain quality of low temperature drying.

Compared with conventional high speed drying, combination drying can reduce energy use up to 50% and high temperature dryer capacity can be doubled or even tripled. Combination drying uses more electrical energy and less fuel, such as LP gas. Consider relative energy costs of electricity and LP gas to determine relative operating costs for both the high temperature and low temperature drying. See Table 3-11 in the evaluating drying systems section for a comparison of relative capacities and costs of high temperature dryers with different cooling methods.

Drying Systems

Low Temperature Bin Dryers

In this book, low temperature drying means slow drying with natural air or air heated up to 10 degrees, Fig 3-10. Drying and storage are in the same bin, so it is sometimes called in-bin drying. The bin has a full perforated floor, high capacity fan or fans, a grain spreader, grain unloading equipment, and possibly a heater.

Low temperature dryers require little grain handling which is a big advantage for this system. The entire drying system (wet holding, drying, and storage) is in the bin, Fig 3-11. Usually when all drying on a farm is done with low temperature drying bins, the whole crop is often stored in one or more bins.

All available research data using low temperature drying is for corn. This data should also be valid for grain sorghum. Low temperature drying can be adapted to most grains, but the discussion in this book relates directly to corn. Contact your state extension agricultural engineer for assistance with using low temperature drying with other grains.

Adequate airflow (cfm/bu) is the key to successful low temperature drying. As discussed in the fan section, drying capacity is directly related to airflow. Adding heat does not reduce the air needed to safely dry, but may cause overdrying. Increasing the airflow dries grain faster and reduces the chance of spoilage.

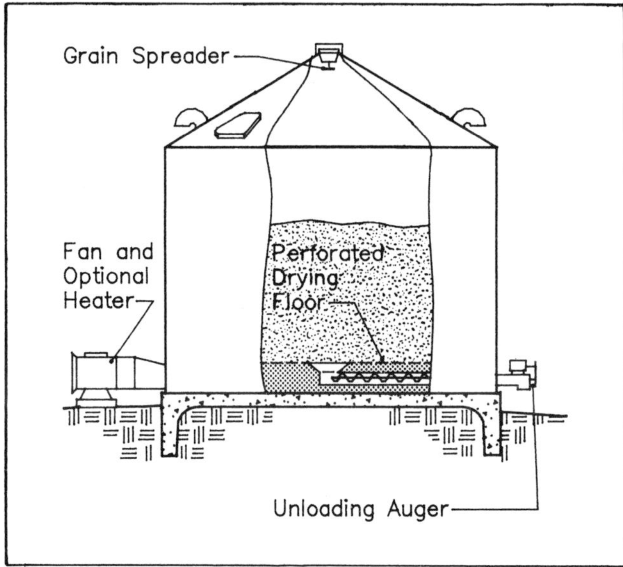

Fig 3-10. Typical low temperature bin dryer.

Fig 3-12 gives recommended design airflow rates (cfm/bu) for selecting low temperature corn drying fans in the North Central region. Increasing airflow rates (cfm/bu) requires larger fans with more power and higher initial and operating costs. But, they can be more reliable.

Three basic procedures for filling a low temperature drying bin are:

- Single filling. Load the bin as rapidly as harvest permits—in one or two days.
- Layer filling. Load the bin with a specific amount (¼, ⅓, or ½ of bin depth) each week.
- Controlled filling. Similar to layer filling, but the amount of grain added each time depends on grain moisture content and drying progress.

All three procedures work if adapted to your specific grain harvesting situation. It is important to thoroughly understand the relationships between airflow, grain and air temperature, grain moisture content, and the grain depth that can be dried without grain spoilage using a low temperature bin dryer. For more information, obtain MWPS-22, *Low Temperature & Solar Grain Drying Handbook*. It discusses these filling procedures in detail.

The maximum moisture content for the single fill low temperature drying method with the recommended airflow is about 22% for corn, 18% for sunflower, 15% for flaxseed, and 17% for small grain. At higher moisture contents, it is usually better to use layer or controlled filling.

Low temperature dryers successfully dry most grains to safe storage moisture contents in the western and west central corn belt without adding heat. These areas may experience possibly 2 "wet years" out of 10, when the relative humidity is 80% or above for a week or more. When this occurs, a temperature rise up to about 10 F helps complete drying in fall. If no heat is added, the only alternative is to cool the grain below 35 F during the winter and complete drying the following spring.

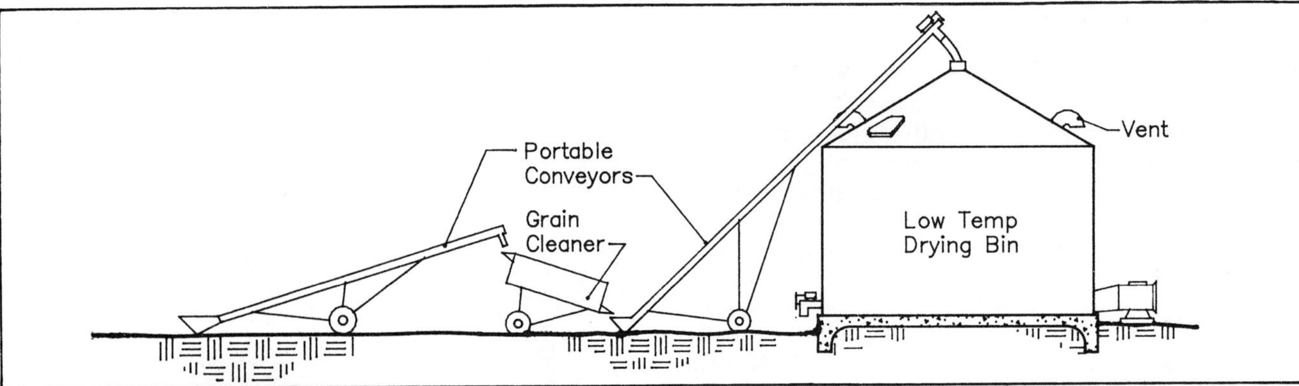

Fig 3-11. Grain handling with a low temperature dryer.

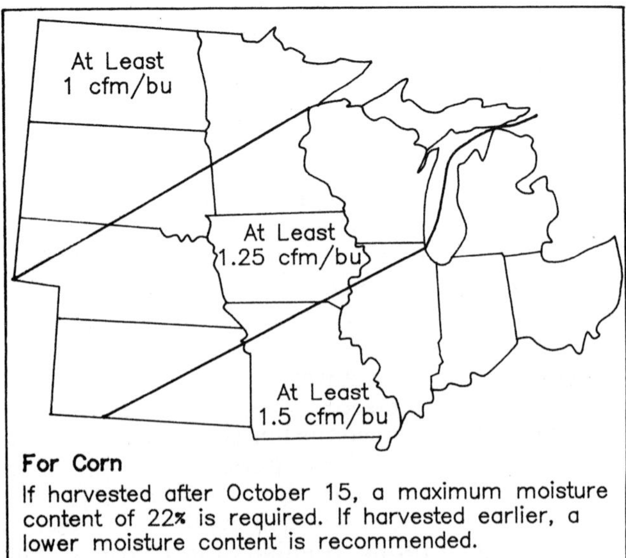

Fig 3-12. Recommended airflow rates for low temperature corn drying.
Use for fan selection. Select a fan to provide at least the airflow rate for your area when the bin is full. These airflow rates were developed by engineers from around the Midwest and published in MWPS-22, *Low Temperature & Solar Grain Drying Handbook.* Contact the Extension Agricultural Engineer in your state for more specific values.

The eastern corn belt has relatively high fall humidities, possibly 5 or more years out of 10. This results in poor field dry down and bin drying. Incoming corn moisture contents may be 25% or more. Heat the drying air up to 10 F under these conditions and use shallow layer or controlled filling or the drying rate will be **very** slow and spoilage may occur. If corn dries in the field to 20%-22%, air temperatures are below 50 F, and grain moisture content in the bin is at least as low as the incoming grain, then the filling procedure may be changed to a single fill full bin method. However, if drying is not complete by mid to late November when air temperatures drop, drying may not be completed until spring.

To finish drying before winter, harvest grain that matures in the fall as early as possible to take advantage of good fall drying weather. However, warm weather can require adjustments in low temperature bin dryer operation. As air temperature increases, grain spoilage rate increases faster than drying rate. So to dry grain safely with the layer or controlled filling methods, shallower layers are required during September and perhaps early October. With the single filling method, wetter grain can be loaded in the bin in October rather than September. Carefully monitor and check low temperature drying bins while drying grain.

Drying small grains in June or July with a low temperature dryer can require severe adjustments in the filling rate, especially in high humidity regions, such as the eastern corn belt. Warm weather with high humidity is ideal for mold growth. Under these conditions, use layer filling with small grains over 17% moisture content after the total accumulated depth reaches 4′.

If heat is used with low temperature drying, common sources are electricity, LP gas, and solar. With solar heat, locate the dryer where adequately sized solar collectors will not interfere with traffic, conveyors, and future growth. MWPS-22 gives details for adding heat, particularly solar, to low temperature drying systems. However, unneeded heat increases drying cost and overdries the bottom grain, causing unnecessary weight loss.

Low temperature drying bins can include stirring equipment for controlling final grain moisture content and increasing airflow rate. Disadvantages are increased investment and maintenance cost and about 1½′ of storage depth lost to the stirring equipment drive mechanism.

Stirring does not significantly increase grain damage. However, fines can sift to the drying floor with over stirring, where they can reduce airflow.

Grain cleaning before drying is recommended for all low temperature drying and is essential for stir drying to minimize fines accumulation on the drying floor. Proper stirring loosens the grain, which increases airflow up to 30%. The additional cost of and storage loss due to stirring equipment in low temperature drying bins can be justified by the advantages of increased airflow and drying capacity in the bin.

Adequate airflow is essential for low temperature drying. Shallower grain depths allow for higher airflow and lower operating costs.

High Temperature Bin Dryers

High temperature bin dryers dry grain with heated air in bin-batch or continuous flow dryers. Two types of bin-batch dryers are on-floor and roof batch dryers.

High temperature bin dryers commonly dry with air heated to between 120 F and 180 F. Grain nutritional value is rarely affected at these temperatures. However, the higher temperatures of this range can affect physical grain quality if delayed cooling is not used; i.e. more stress cracks and kernel brittleness, less color, and reduced germination. Temperatures above 120 F for malting barley or 150 F for wheat can affect marketability. Keep drying air below 110 F for seed grain.

Drying and storage are usually in separate bins. The drying bin has a perforated floor, drying fan, heater, grain spreader, and grain unloading equipment. Batch drying bins may also have grain stirring equipment. One drying bin can dry grain for one or more storage bins.

High temperature bin dryers are primarily for drying grain, but can be used to store the last batch of grain dried. Some storage problems can occur with storing grain in the drying bin. Conveying grain to storage can reduce storage problems by mixing drier and wetter grain. If the last batch is stored in the drying bin, mixing does not occur. To reduce storage problems, consider the following procedures.

- Select shorter sidewall drying bins to minimize the potential for problem grain.
- Install a perforated wall liner or perforated tubes around the inner wall to minimize sidewall moisture condensation. See the section on condensation problems for installation details.
- Dry the last batch with unheated air to minimize uneven drying. If grain is dried during late fall, drying may need to be completed in the spring.
- If grain moisture content is less than 18%, cool the grain and maintain the temperature below 40 F and market or feed the grain before spring.

On-floor bin-batch dryer

An on-floor bin-batch dryer dries a layer of grain on a perforated floor. Limit batch drying depth so the moisture content of the dried batch varies less than about 5 percentage points (e.g. 16%-11%) between top and bottom. Moving the dried batch to storage mixes the wetter top grain with the drier bottom grain for safer storage. As grain depth increases, drying capacity decreases and it is more difficult to manage the drying process.

The bins can be equipped with or without stirring equipment. Recommended grain depth for drying is 2½′-4′ without stirring, Fig 3-13. With stirring equipment, grain depth for drying can be up to about 9′, Fig 3-14. Proper management practices are necessary for successful drying at greater depths. Stirring equipment allows greater depths without excessive moisture differences between top and bottom grain. Table 3-8 shows batch sizes for several bin diameters.

With an on-floor bin-batch dryer, grain is usually cooled immediately in the dryer to minimize the time grain is in the bin. An alternate delayed cooling method with bin-batch dryers is to turn the fan off after drying for at least 4 hrs to allow the grain to steep before cooling. Then turn the fan on again to cool and finish drying the grain. See the cooling section.

Fig 3-15 shows one of several effective bin arrangements.

For bins without stirring, select fan capacity, drying air temperature, and grain depth so at least one batch can be harvested, dried, cooled, and moved to storage each day. With convenient wet grain holding and fast handling equipment, it is possible to dry two or three batches per day.

If drying only one batch a day, you can load wet grain directly into the drying bin. For more than one batch per day, hold wet grain in wagons or trucks. A holding bin with enough fan capacity to cool wet grain and prevent heating provides flexibility to hold grain for several days. See the section on wet grain

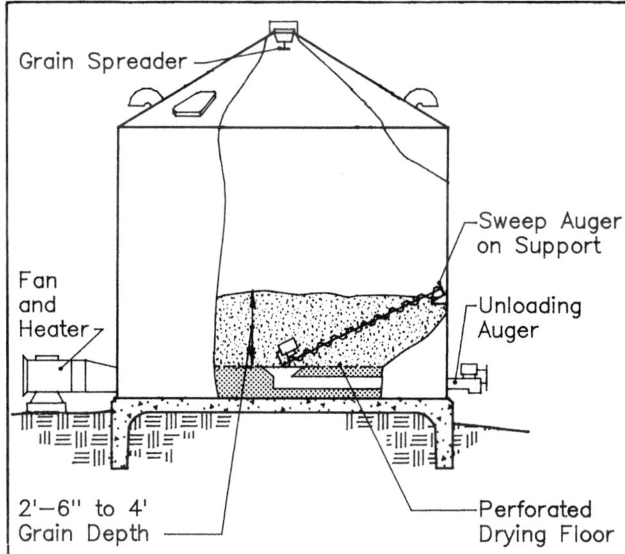

Fig 3-13. Batch-bin dryer without stirring.

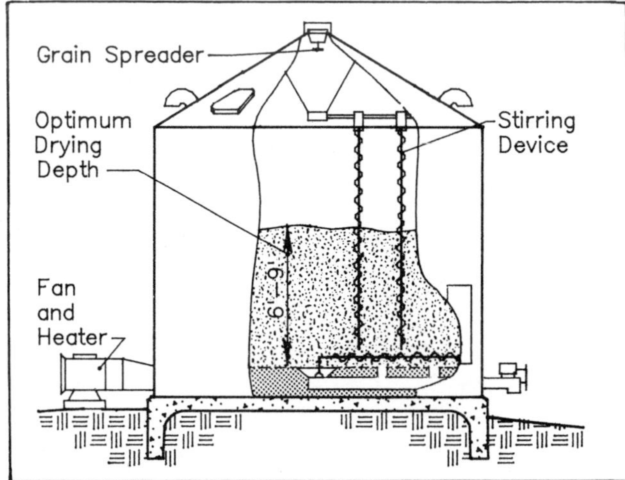

Fig 3-14. Batch-bin dryer with stirring.

holding. Transferring grain from wet holding to the dryer can be controlled automatically. Batch drying requires a high capacity conveyor to minimize loading and unloading time and maximize drying capacity.

Low capacity, inconvenient grain unloading equipment often slows bin-batch drying. Regardless of batch size and drying and cooling procedures, provide **minimum** conveying capacity of at least 1,000 bu/hr; up to 2,500 bu/hr is preferred. A permanent gear-driven sweep auger eliminates moving a motor-driven sweep auger in and out of the bin to unload each batch. Another possibility is to install an underfloor auger with multiple inlets, so a motor-driven sweep stays in place between batches, Fig 3-14. Place the sweep directly over the multiple inlets before loading the batch. Grain flowing into the inlets reduces grain depth over the sweep and permits it to be started. If stirring equipment is used, be sure stirrers do not hit sweep augers or motor.

Consider sweep augers with back boards or shields behind the auger to more effectively remove grain and help reduce fines buildup on the perforated floor. Accumulated fines reduce airflow and drying capacity and can cause bin fires. After every three or four batches, remove fines from the perforated floor to assure proper air movement and reduce fire risk.

Compared with other drying systems, bin-batch dryers often cost less per unit of drying capacity, but require more labor. Someone usually needs to supervise the transfer of each batch to storage, fill for the next batch, and level grain if spreaders do not operate properly.

Table 3-8. Bushels per batch in bin dryers.

Bin dia. ft	bu/ft of depth	Approximate batch size Bin depth, ft 2½[a]	4[a]	6[b]	9[b]
		- - - - - - - - bu	- - - - - - - -		
18	204	510	816	1,224	1,836
21	277	693	1,108	1,662	2,493
24	362	905	1,448	2,172	3,258
27	458	1,145	1,832	2,748	4,122
30	565	1,413	2,260	3,390	5,085

[a]Stirring not required.
[b]Stirring required.

Consider a 24 hr time clock for convenient bin batch dryer control. Set the timer to turn the burner off when the batch is estimated to be dry. The fan continues to run and cool the grain for transfer to storage.

Roof bin-batch dryer

Roof bin-batch dryers dry a layer of grain about 2½′ deep on a coned perforated floor under the roof, Fig 3-17. The approximate batch size for different diameter bins is shown in Table 3-9. Wet grain is loaded onto the coned perforated floor for drying. Hot dried grain drops onto a totally or partially perforated floor for cooling and/or storage. Hot grain is cooled with a separate fan while the drying chamber is refilled with wet grain and drying starts again. Up to four batches can be dried per day.

The space between the grain being cooled and the drying floor is a hot air plenum. Heat from the cooling grain preheats some of the drying air.

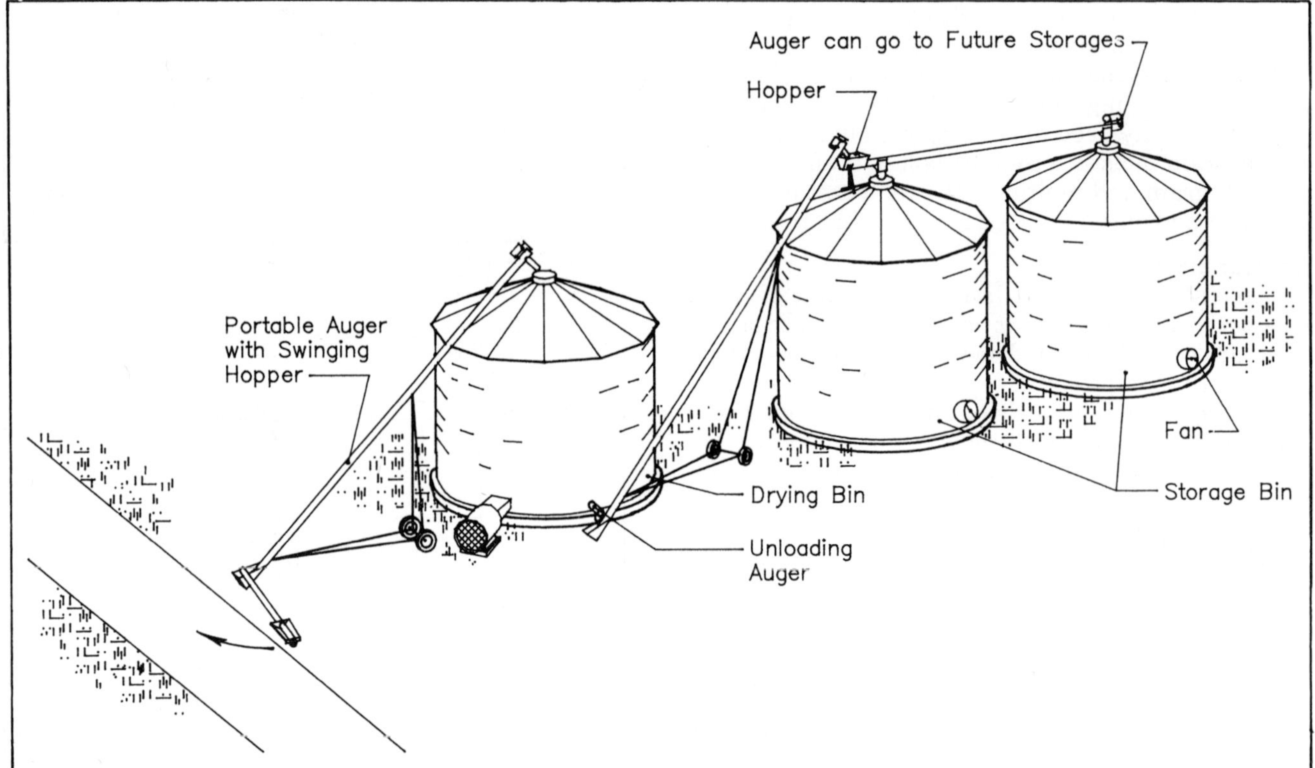

Fig 3-15. High temperature bin-batch drying system.
Load dryer with portable auger. Another portable auger conveys dry grain to the overhead auger.

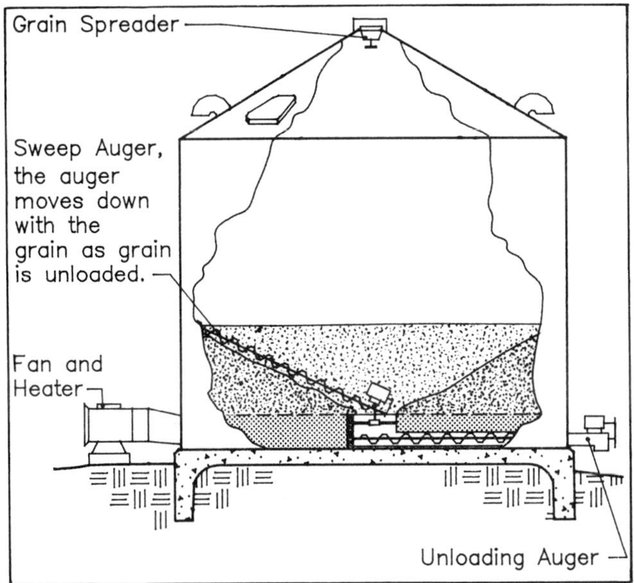

Fig 3-16. Gear-driven sweep auger.

Table 3-9. Roof bin-batch dryer drying section capacity.
Values are approximate.

Bin dia. ft	Batch size bu
18	540
21	750
24	1,000
27	1,250
30	1,500
36	2,100

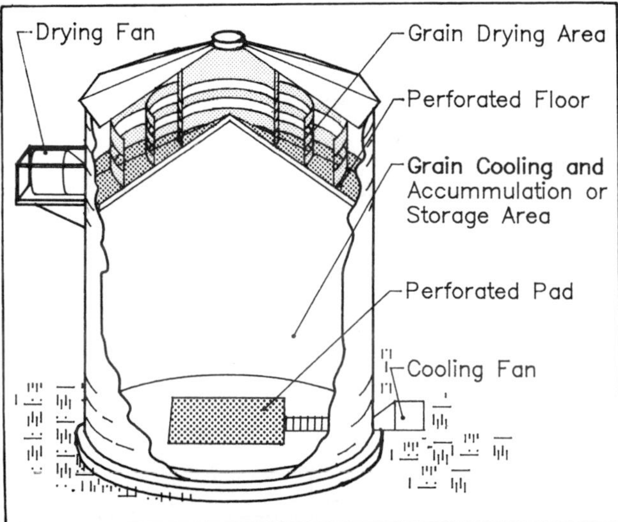

Fig 3-17. Roof bin-batch dryer.
The drying capacity of a roof bin-batch dryer can dry grain for several dry grain storage bins which improves effectiveness and reduces system cost.

There are few bottlenecks in handling dried grain with roof bin-batch dryers. An entire batch can be dumped from the drying area to the cooling area in less than 1 min. However, loading requires more time. Use high capacity conveying equipment to minimize delays. Automatically controlled low capacity conveyors can be used to move dried, cooled grain to storage.

Continuous flow bin dryer

Continuous flow bin dryers are usually a bin with a perforated drying floor, fan, heater, grain spreader, grain unloading equipment and an auger to transfer grain to storage. These dryers unload hot, dried grain intermittently from the drying bin to bins equipped for in-bin cooling and storage.

Fig 3-18 shows a portable inclined auger to transfer grain to storage. A recirculating and transfer auger in the bin is shown in Fig 3-19. Grain flow to storage and cooling bins is automatically controlled. Size conveyors to meet the maximum expected drying capacity.

Grain recirculation or stirring equipment is used only when the last batch of grain is dried and stored in the drying bin. If grain is stored in the drying bin, a perforated wall liner or evenly spaced perforated tubes are needed to minimize condensation.

Every few days, completely empty the drying bin and remove accumulated fines from the perforated floor to reduce airflow blockage and potential bin fires. Check the plenum under the drying floor every few years for fines accumulation. When someone is in the bin, make sure the sweep, stirring, and unloading augers cannot be started.

Separate wet holding bins for high temperature bin dryers give maximum flexibility to maintain optimum drying depth. With separate wet holding, shallower drying depths allow for increased airflow and drying rates to maintain higher drying capacity during harvest.

Grain conveying from wet holding to drying can be controlled automatically. Provide conveying capacity equal to at least the maximum drying rate. Older, smaller drying bins with a perforated floor and fan are often used for wet grain holding. See the section on wet grain holding.

Drying capacity.

Drying capacity for all high temperature bin dryers varies considerably with grain type, initial and final grain moisture content, loading and unloading time, airflow, drying air temperature, outdoor air conditions, and whether the grain is cooled in the bin. Changing any of these factors can either increase or decrease the drying capacity of the dryer.

The specific conveying equipment used can affect the amount of time required for loading and unloading the dryer. Decreasing the loading and unloading time reduces delays and increases available drying time and drying capacity.

Continuous flow bin dryers require little delay time for conveying because grain is loaded and unloaded as drying continues. Bin-batch dryers require more time for conveying grain which reduces available drying time. A roof bin-batch dryer can unload dried grain in less than a minute, but requires more time for loading. Minimize loading and unloading delays in roof and on-floor bin-batch dryers with high capacity conveying equipment.

As discussed in the fan section, relative drying capacity (bu/hr) is directly proportional to airflow (cfm/bu). As airflow increases, drying capacity in-

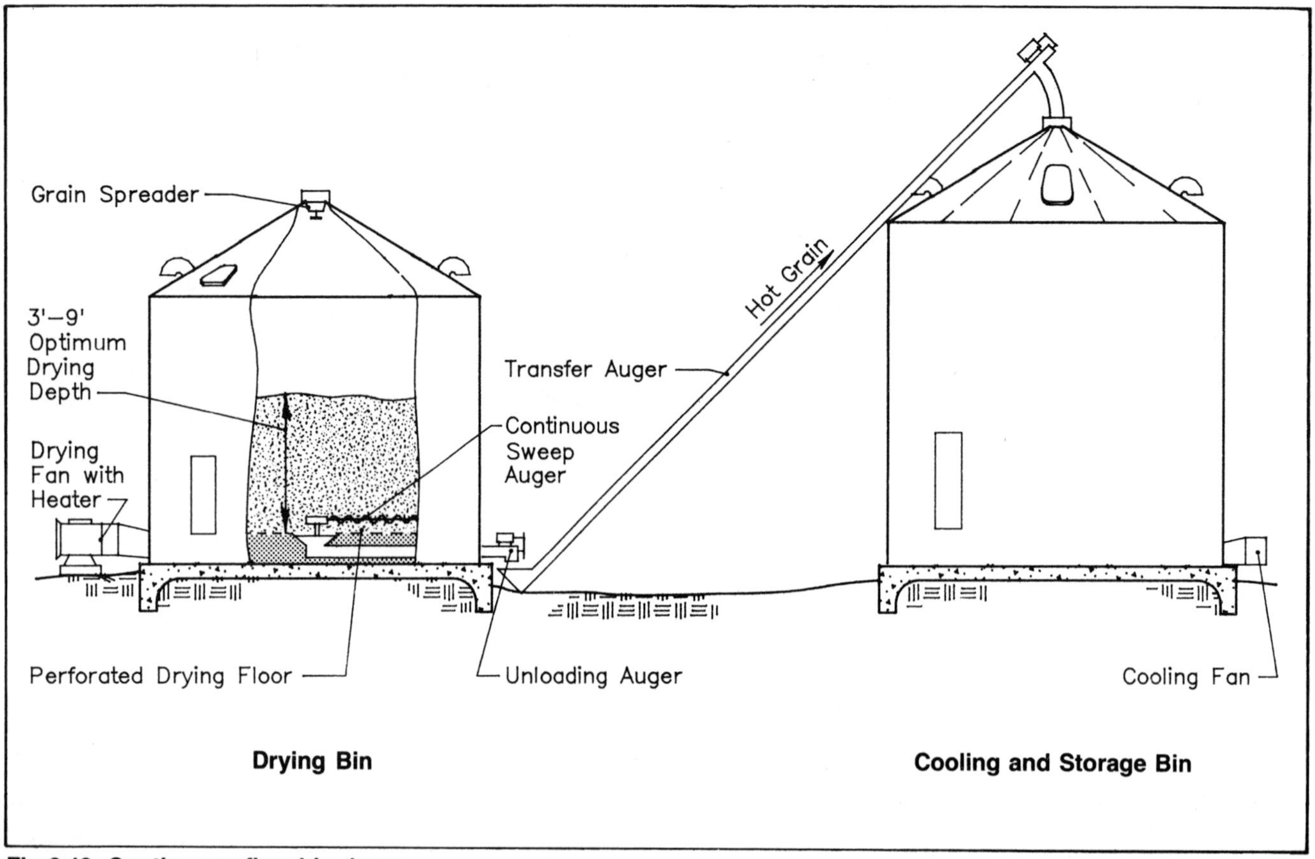

Grain Spreader
3'–9' Optimum Drying Depth
Drying Fan with Heater
Transfer Auger
Hot Grain
Continuous Sweep Auger
Perforated Drying Floor
Unloading Auger
Cooling Fan
Drying Bin
Cooling and Storage Bin

Fig 3-18. Continuous flow bin dryer.
A transfer auger moves grain intermittently to cooling in storage bins.

Grain Recirculator, for the last batch of the season
Grain Spreader
Hot Grain
Transfer Auger
3'–9' Optimum Drying Depth
Drying Fan with Heater
Continuous Sweep Auger
Perforated Drying Floor
Unloading Auger
Cooling Fan
Drying Bin
Cooling and Storage Bin

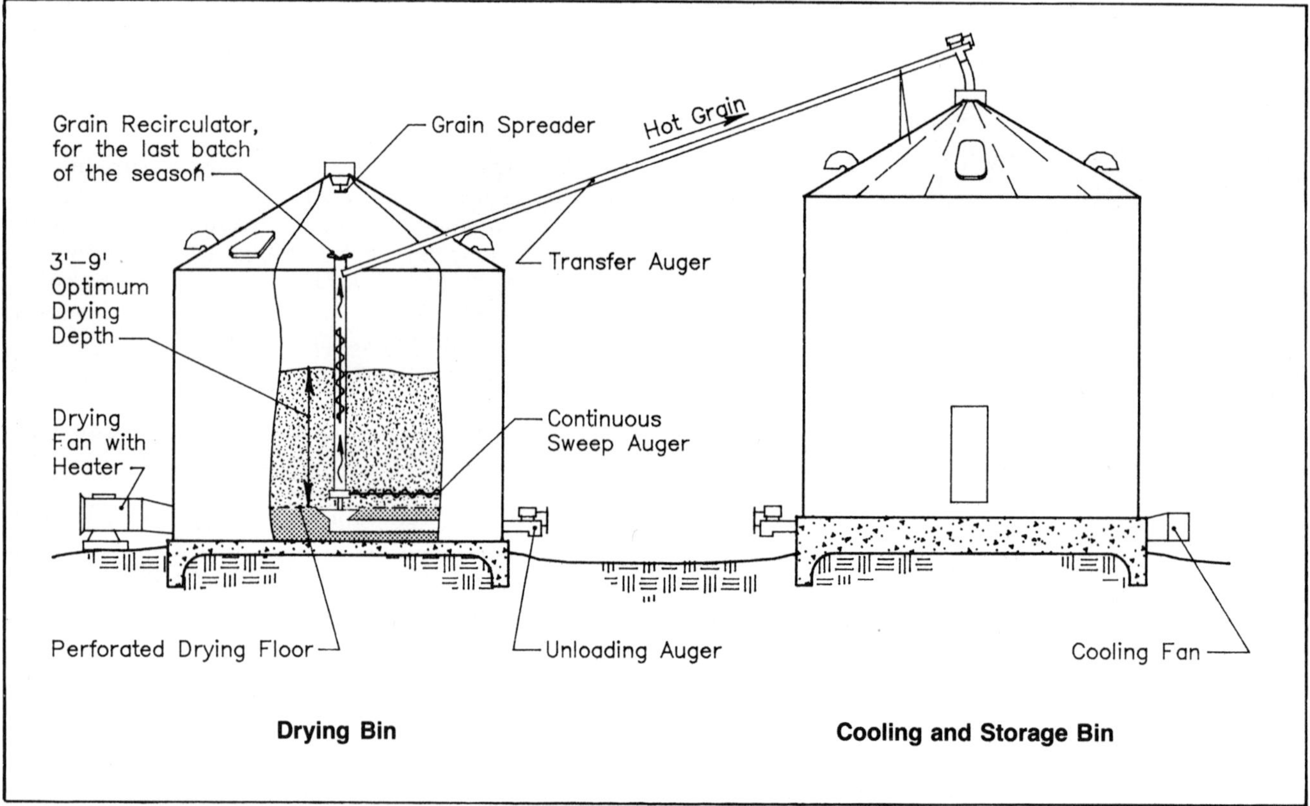

Fig 3-19. Continuous flow recirculating bin dryer.
At least one more storage bin equipped to properly cool hot grain is common.

creases. Deeper grain depths increase airflow resistance which reduces the amount of air delivered to the bin decreasing drying capacity. Consider high temperature drying bins with shorter walls than the storage bins, Fig 3-20. Replace lost storage capacity in the drying bin with taller storage bins.

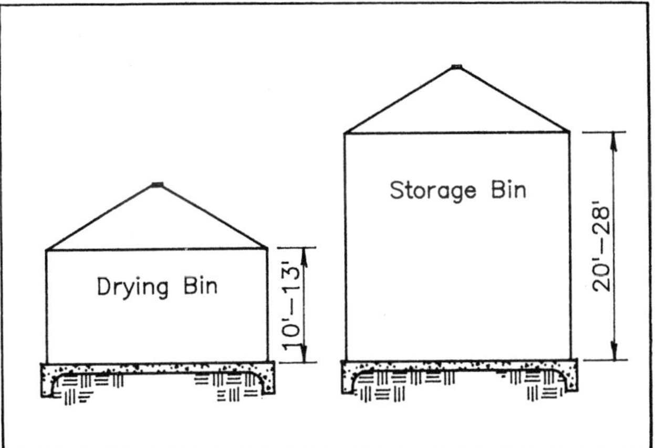

Fig 3-20. Recommended bin sizes for drying and storage.

Roof bin-batch dryers dry grain in batches 2½′ deep. On-floor bin-batch dryers without stirring often dry grain in batches up to 4′ deep. Continuous flow and on-floor bin-batch dryers with stirring can be operated with grain depths of 9′ or greater, but have higher drying capacities at shallower depths. Drying depths up to 6′ are preferred with continuous flow and on-floor bin-batch dryers with stirring.

In addition to using shallower grain depths to increase drying capacity, adding another fan and heater can increase airflow and drying capacity. See Table 3-4. If more than one fan and heater are installed on a bin, use the same type, size, and brand to minimize uneven air distribution and drying.

Increasing drying air temperature also increases drying capacity. Tables 3-5 and 3-6 show the effect drying air temperature has on drying capacity. Evaluate your situation and select the drying air temperature that fits your needs. Consider grain type, available equipment, operating procedures, and previous experience with high temperature drying.

Consider using delayed cooling methods to increase drying capacity and reduce drying costs and stress cracks. See the grain cooling section.

Table 3-10 shows the maximum potential drying capacity for high temperature bin dryers with grain depths up to 4′. This table assumes minimum grain loading and unloading time, maximum practical fan capacity, 180 F drying air, and no grain cooling in the drying bin. These values are the absolute maximum practical drying capacities under perfect conditions. It is impractical to expect these airflow rates and drying capacities at grain depths greater than 4′. You should not expect your drying system to achieve these capacities. They are presented as examples for comparison between different bin diameters.

Table 3-10. Maximum practical high temperature bin dryer capacities.
These are estimated capacities to dry shelled corn from 25½% to 15½% including moisture removed in cooling. Values assume maximum practical drying fan airflow and 180 F drying air temperature.

Bin dia. ft	Airflow cfm	Drying capacity bu/hr	Drying capacity bu/24 hr
18	12,000	83	2,000
21	16,500	117	2,800
24	21,250	150	3,600
27	27,250	192	4,600
30	33,750	238	5,700
33	40,250	283	6,800
36	48,000	338	8,100
40	59,000	417	10,000

Most high temperature drying bins are not selected or equipped for the maximum potential drying capacity. Many high temperature bin dryers are selected to provide a drying capacity of about ½ the capacity shown in Table 3-10. When selecting drying bin capacity consider cost, required capacity, operating requirements, and potential for future expansion.

Safety

Fires can occur in high temperature bin dryers and are more common with some grains, especially sunflower and grain sorghum. Fans and burners require safety controls such as thermostats, high temperature limit switches, airflow switches, and flame detectors. Control burners on high temperature dryers with thermostats. Check controls periodically for proper operation. **Constantly** supervise sunflower and grain sorghum drying at high temperatures. Fires can result from trash sucked into the fan and blown through the flame. Clean trash around and in the dryer daily. Consider building a vertical air intake around the fan and/or ducting the fan intake away from the dryer. Size the duct for three times the fan's cross-sectional area.

High Temperature Self-Contained Dryers

High temperature self-contained dryers can be manual batch, automatic batch, or continuous flow. All drying equipment is built into the dryer by the manufacturer. Filling and unloading equipment may or may not be built into the dryer.

Most self-contained high temperature dryers are movable and some are used as portable dryers. With a portable dryer, fuel, electricity, and grain handling equipment are required at each drying site.

Regardless of dryer portability, a complete system also includes wet holding and cooling. These dryers are built to cool grain in the dryer. However, they can be adapted to eliminate cooling in the dryer and move hot grain to a separate bin for cooling. Delayed cooling outside the dryer increases dryer capacity, reduces drying cost, and reduces stress cracks. Wet grain holding and cooling techniques differ between dryer types, so drying capacity and grain flow needs vary. See the grain cooling section.

Dryer selection

Consider filling, drying, and unloading requirements when selecting a high temperature self-contained dryer. Dryer type affects the selection of wet holding bins, conveying equipment, and grain cooling method. Drying speed, labor input, and required supervision during operation are specific to each dryer type. Consider your situation, available labor, and available conveying, wet grain holding, and grain cooling equipment before selecting a high temperature self-contained dryer.

Manual batch dryer

These dryers are typically selected for drying relatively small volumes of grain per year. An advantage of manual batch dryers is portability. They can be used in several locations, including in-field drying. Only the dryer, LP gas tank, and perhaps one conveyor need to be moved. Disadvantages are more labor and supervision are needed for loading, drying, and unloading each batch.

These dryers require manual loading and unloading and are often powered with a tractor PTO, Fig 3-21. Protect the tractor with safety controls to turn off the tractor in case of low oil pressure, high engine temperature, or low engine rpm.

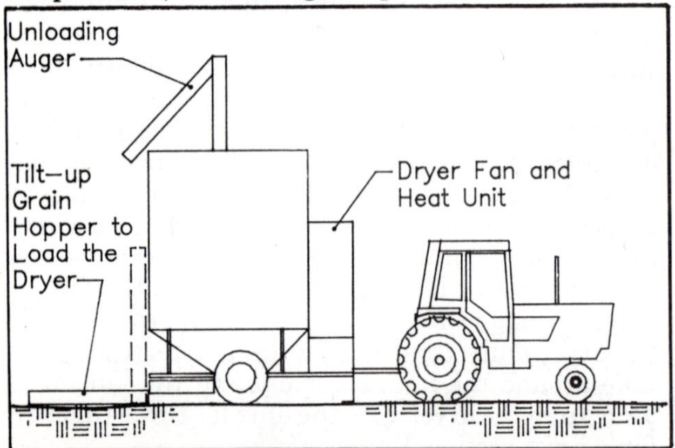

Fig 3-21. Manual batch dryer.
Typically portable, PTO powered, and with self-contained grain handling.

Automatic batch dryers

Automatic batch dryers operate similar to manual batch dryers, except loading and unloading are automatically controlled, Fig 3-22. Batch dryers do not dry grain during filling and unloading, so minimize downtime with high capacity grain conveyors. A gravity flow wet holding bin is often installed over the dryer to provide fast loading. Use a high capacity conveyor, such as a bucket elevator, for fast dryer unloading.

Consider delayed grain cooling to increase dryer capacity and reduce drying costs and stress cracks.

Continuous flow dryers

These dryers are loaded and unloaded either continuously or in frequent intermittent cycles. Size the loading and unloading conveyors for at least the **maximum** grain drying capacity. Maximum drying capacity is typically less than 500 bu/hr, so wet and dry conveying capacity requirements are low.

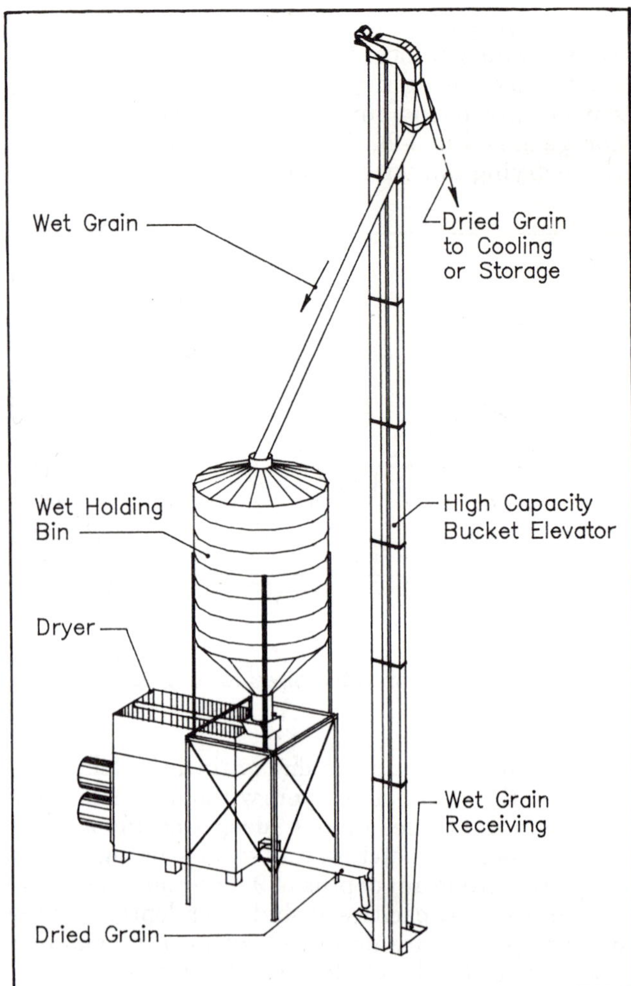

Fig 3-22. Automatic batch drying system.

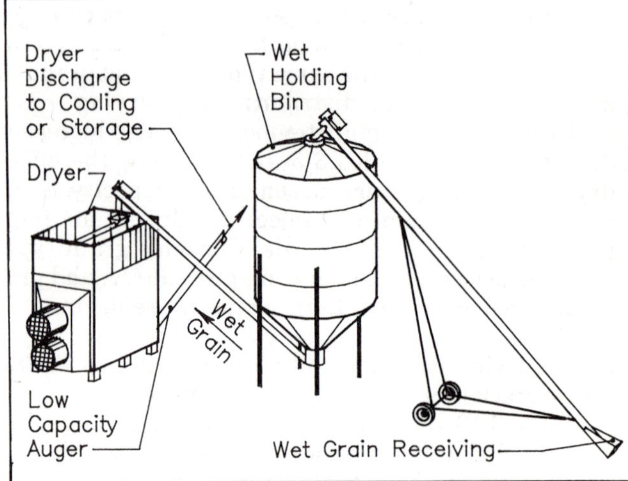

Fig 3-23. Typical continuous flow drying system.
With ground level wet grain holding bin.

A ground level wet holding bin with an auger to fill the dryer is common. Low capacity conveyors, such as augers, can be used to move dried grain to cooling or storage bins. Fig 3-23 shows a typical continuous flow drying system. Even for low capacity conveying, wet grain augers, high horsepower, and at least double V-belt drives are required.

Drying capacity

High temperature dryer capacities are often quoted for ideal drying conditions. However, ideal drying conditions rarely occur on the farm. The realistic capacity is often 70%-80% of the advertised value.

High temperature, self-contained dryers require an operating wet holding bin for maximum performance. Extra wet holding, often in another bin, allows for more flexibility and increased drying capacity. See the holding wet grain section to size operating wet holding capacity. You may not need as much operating capacity at first, but it may be difficult and expensive to increase later if you dry more hours per day or buy a higher capacity dryer.

A well designed system can dry 24 hr/day, but many farmers operate 10 to 16 hr, so they can supervise drying, particularly with widely varying grain moisture content. High temperature sunflower and grain sorghum drying requires **constant** supervision because of potential fire hazards.

To estimate required drying capacity, divide the maximum expected bushels harvested/day by hours of dryer operation per day. For example, 45,000 bu harvested in 15 days, requires a dryer capacity of 3,000 bu/day (45,000 bu ÷ 15 days). To operate the dryer only 10 hr/day, dryer capacity must be 300 bu/hr (3,000 bu/day ÷ 10 hr/day). To operate 15 hr/day, capacity can be 200 bu/hr. If the dryer runs automatically 24 hr/day, 125 bu/hr capacity is adequate.

Plan the drying system so annual drying capacity can be increased at least 50% (multiply by 1.5). For more flexibility, plan for double the drying capacity. This provides the opportunity to evaluate what might change within the probable facility life. Drying capacity can be expanded by: drying more hours per day, dryeration, in-bin cooling, combination drying, or a combination of these. A 300 bu/hr dryer with an in-dryer cooler operating 12 hr/day dries 3,600 bu/day. Drying 20 hr/day increases capacity to 6,000 bu/day. Converting the dryer to full heat operation with dryeration increases capacity to 5,800 bu/day with 12 hr drying or 9,600 bu/day with 20 hr drying. If the system includes low temperature drying bins (combination drying), dryer capacity goes to 9,000 bu in 12 hr or 15,000 bu in 20 hr.

Energy conservation

Grain drying consumes a lot of energy, so save what you can. For example, dryeration, one method of delayed cooling, can cut 25% off energy consumption of high temperature drying while increasing drying capacity up to 65% for 10 percentage points of moisture removed. In-bin cooling and combination drying with high temperature dryers are other cooling methods that can also save energy. See the section on grain cooling.

Recovering some discharge air from column type continuous flow dryers can save up to 30% energy. This much savings usually only occurs when drying relatively dry grain during cold weather. Some dryers recover and recirculate all discharge cooling air and others recover all cooling air and some of the discharge air from the lower portions of the heated air section. Good air recovery systems are practical energy savers with very little loss in dryer capacity. Several manufacturers have energy conservation kits available for older model dryers.

Recovering discharge air from the lower part of a full heat (no in-dryer cooling) continuous flow dryer adds up to 20% more energy savings. Actual energy savings by air recovery depends on dryer airflow (cfm/bu), final grain moisture content, moisture content reduction in the dryer, outdoor and drying air temperatures, and the proportion of total dryer air flow that is recovered.

Evaluating Drying Systems

Relative Costs and Capacities

Capacity and operating costs of high temperature drying depend on the grain cooling method: cooling in dryer, in-bin cooling, dryeration, or combination high and low temperature drying. See Table 3-11.

Table 3-11. Estimated relative operating costs.
For drying/cooling corn from 25.5% to 15.5%.

High temperature drying with:	Relative capacity bu/time	Gallons propane/100 bu	Electrical energy Kw-hr/100 bu
In-dryer cooling	1	20	10
In-bin cooling	1.35	17.5	8
Dryeration	1.6	14.5	7
Combination drying	2.5*	8	70-110

*High temperature drying capacity.

Delayed cooling increases heated air dryer capacity, lowers dryer operating cost, and reduces stress cracks in grain. Delayed cooling increases initial cost of drying, cooling, and grain handling equipment.

Example 3-2:

Determine the relative capacities and operating costs if a high temperature dryer is converted to a delayed cooling system. The high temperature dryer originally dries and cools 200 bu/hr. Assume energy costs are 70¢/gal for propane and 10¢/kw-hr for electricity.

Solution:

1. With in-dryer cooling
   The drying capacity is 200 bu/hr for a high temperature dryer with in-dryer cooling.

   From Table 3-11, the high temperature dryer uses 20 gal of propane and 10 kw-hr of electricity/100 bu of corn dried from 25.5% to 15.5%. Dryer operating cost is:

   70¢/gal × 20 gal/100 bu
   + 10¢/kw-hr × 10 kw-hr/100 bu = 15¢/bu

2. With in-bin cooling
   From Table 3-11, the relative drying capacity is 1.35 and the dryer uses 17.5 gal of propane and 8 kw-hr of electricity/100 bu of corn. Drying capacity increases to:

   1.35 × 200 bu/hr = 270 bu/hr

   and operating costs decrease to:

   70¢/gal × 17.5 gal/100 bu
   + 10¢/kw-hr × 8 kw-hr/100 bu = 13.1¢/bu

3. With dryeration
   From Table 3-11, the relative drying capacity is 1.6 and the dryer uses 14.5 gal of propane and 7 kw-hr of electricity/100 bu of corn. The drying capacity increases to:

   1.6 × 200 bu/hr = 320 bu/hr

   and operating costs are:
   70¢/gal × 14.5 gal/100 bu
   + 10¢/kw-hr × 7 kw-hr/100 bu = 10.9¢/bu

4. With combination high/low temperature drying
   From Table 3-11, the relative high temperature drying capacity is 2.5 and the dryer uses 8 gal of propane and 90 kw-hr of electricity/100 bu of corn. The drying capacity increases to: 2.5 × 200 bu/hr = 500 bu/hr

   and operating costs decrease compared to in-dryer cooling to:

   70¢/gal × 8 gal/100 bu
   + 10¢/kw-hr × 90 kw-hr/100 bu = 14.6¢/bu

Table 3-12. Summary of Example 3-2.
Values are relative capacities and operating costs from Example 3-2 of operating a high temperature dryer with cooling in dryer, in-bin cooling, dryeration, and combination high/low dryer.

High temperature drying, with:	Estimated drying capacity bu/hr	Estimated fuel operating cost, ¢/bu
Cooling in dryer	200	15.0
In-bin cooling	270	13.1
Dryeration	320	10.9
Combination drying	500*	14.6

*High temperature drying capacity.

The examples summarized in Table 3-12 include only drying capacity and energy costs. To compare drying systems consider all equipment costs. Compare equipment investments for wet holding, drying, cooling, and handling for each drying system. Drying equipment dealers and other farmers with similar equipment often have this information.

From investment cost, estimate annual fixed cost (depreciation, interest, repairs, taxes) at 18%-20% of equipment first cost. Consult your extension farm management specialist to help evaluate the cost effectiveness of your planned system.

4. STORAGE

Storage does not improve grain quality. The purpose of good storage is to maintain grain quality after it is harvested by protecting it from weather, rodents, and insects. Generally, to maintain grain quality, all biological and insect activity must be controlled.

Several types of storage can be used. The type selected, such as permanent or temporary, depends on your individual needs. Regardless of the type, proper selection, sizing, and location are essential to a successful grain storage system.

Selecting Grain Storage

Management flexibility is related to size and number of storages. If only a single large storage is available, flexibility is very limited. For example, with only one storage, if some grain had to be stored a second year, the risk of an insect infestation problem in the new grain increases. Other factors that affect storage selection decisions are crop production ratios (e.g. corn to soybeans, wheat to grain sorghum, sunflower to barley) may change or a given variety may show signs of field mold and need isolation.

As a general rule, never have more than half of the total annual grain production of one grain type in a single storage. A mixture of storage sizes is best. Different size storages allow more flexibility to meet changing needs from year to year.

Grain is generally easier to manage in small storages than in large storages. In small storages, the grain mass influencing cooling, heating, and moisture migration is less and grain is easier to probe and check. and check.

To optimize grain storage costs, management flexibility, and production/marketing options, choose storage sizes large enough to meet your storage needs at a reasonable cost per bushel.

Metal Bins

Round metal bins are the most common grain storage structure. They are convenient to aerate, load, and unload and are relatively low cost, compact, and fit well into a total system. However, round bins have few alternate uses. A wide range of grain handling, conditioning, and safety equipment is available to use with round bins.

When selecting storage bins there are no set rules for determining the number and size. Choose bins that best fit your need now and provide flexibility in the future. Select bins large enough to meet your storage needs at a reasonable cost per bushel. As bin size increases from 1,000 bu to 12,000 bu, the cost per bushel drops rapidly. For 12,000 bu to 100,000 bu and larger bins, the cost per bushel may still decrease, but not as rapidly. The cost per bushel may even increase with larger bins because of increased cost of trussed roofs, taller sidewalls, etc. The cost per bushel difference between two 10,000 bu bins and one 20,000 bu bin may be greater, but this is a **one-time cost.** The flexibility of two smaller bins rather than one large bin can return the difference in cost because of more effective grain management and/or marketing.

Different handling equipment is needed for two smaller bins than for one large bin. Larger bins are usually taller, which may require a longer conveyor or a roof auger. Filling two smaller bins will require moving a conveyor, a roof-top conveyor, or two gravity spouts. Unloading may require two underfloor augers. However, the cost of this additional equipment is small when considered over the life of the bin—20 years or longer.

Flat bottom bins are usually filled only once per season. Round bins can be unloaded with underfloor augers emptying from a sump in the center of the floor. Plan extra underfloor auger intakes for flexibility and convenience. Consider an additional auger intake about 2′ from the center sump and another near the bin door, Fig 4-1. The intake 2′ from the center can be used in an emergency if the center intake plugs. An intake near the door is for convenience to move grain away from the door when unloading. Sweep augers minimize labor to remove all of the grain from the bin.

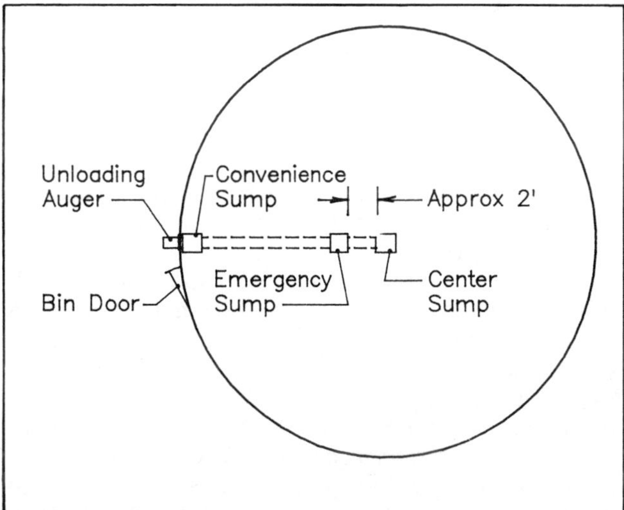

Fig 4-1. Round bin auger intakes.

Fully perforated floors provide maximum flexibility in a total grain system. They can be use for drying, cooling, and wet grain holding, but are more expensive for aeration systems. Duct or aeration pad air distribution systems are often selected because of lower costs. A properly designed and installed duct system can provide adequate air distribution.

Proper bin foundation and floor construction is important—follow manufacturers' specifications. The bin floor should be level. Any irregularities in the floor surface at or outside the bin wall that can collect water can cause serious problems. If bin size exceeds 50,000 bu, obtain a soil bearing test to determine adequate footing and floor design. Drainage away from the bin is needed to prevent moisture buildup on the floor and around the bin. Construct the bin floor at least 12″ above grade to minimize surface water problems and provide adequate clearance for the unloading auger, Fig 4-2.

Place a vapor barrier, usually at least a 4 mil plastic sheet on sand fill, before placing the concrete floor. Seal the connection between the bin wall and concrete floor or foundation to prevent moisture and air from passing through. Most commercial bin manufacturers provide a mastic seal to place between the bin wall and floor, Fig 4-2.

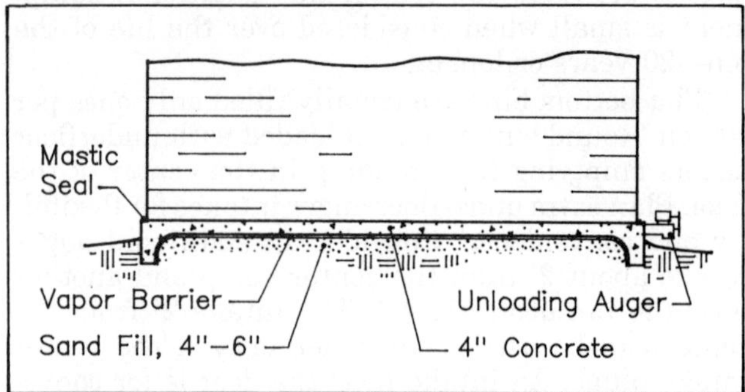

Fig 4-2. Round bin wall and floor seal.

Consider safety precautions and practices when installing round metal bins. Safety practices include installing caged ladders, stairs, hand rails, and roof ladders on the outside of bins, following OSHA regulations, and keeping absolutely everyone out of the bin during loading and unloading.

Capacity

Round metal bins are typically 15′-60′ in diameter. The volume capacity of round bins can be determined using the definition of a standard bushel. A standard volume bushel equals 1.245 ft^3 per bushel. Table 6-5 gives bin capacities in bushels by volume based on a standard bushel for most bin diameters. The capacity of a level filled round bin can be estimated using Eq 4-1. This equation is based on an approximate conversion factor of 0.8 bu/ft^3. This conversion factor is determined by rounding off the standard volume bushel to 1.25 ft^3/bu. To get the commonly used ft^3/bu conversion factor divide 1 by 1.25 ft^3/bu to get 0.8 bu/ft^3.

Eq 4-1.

$$RBC = 0.785 \times D \times D \times H \times 0.8$$

RBC = estimated capacity of level filled round bin, bu
0.785 = constant
D = bin diameter, ft
H = bin height, ft
0.8 = conversion factor, ft^3 to bu, bu/ft^3

For example, a 30′ diameter bin filled to a level depth of 18′ holds an estimated 10,174 bu (0.785 × 30′ × 30′ × 18′ × 0.8). Table 6-5 values may differ from Eq 4-1 because of the approximate bu/ft^3 conversion factor. From Table 6-5, a 30′ diameter bin filled to a uniform depth of 18′ contains 10,220 bu by volume. However, bushel capacities by weight may differ from the values in Table 6-5 because of compaction, test weight, and distribution of fines in the void space between the kernels. If each 1.245 ft^3 (one bushel by volume) of corn in the bin weighed 56 lb (one bushel of No. 2 corn by weight), the bin would contain 10,220 bu. The only accurate method of determining the capacity of a bin is to weigh the grain either loaded into or out of the bin.

A factor of 5% is commonly used to account for grain compaction. With a 5% compaction factor, the 30′ diameter bin 18′ level full of corn would contain 10,731 bu (10,220 bu × 1.05) at 56 lb test weight. A higher or lower test weight would increase or decrease the capacity proportionately. A grain spreader tends to increase this factor. A high percentage of fines increases the capacity because of the weight of fines in the space between kernels.

Flat Storage

Flat storage structures are common for storing large quantities of a single grain. Usually, the storage is filled and unloaded only once a year because loading and unloading flat storages is not as convenient as with round bins. Flat storages are suitable for grain that does not need to be dried before storage. Many farmers and grain elevators, especially in the western plains states where harvested grain requires little or no drying, select flat grain storages because of more flexibility.

Flat grain storage offers the possibility of alternative uses, such as machinery storage, hay storage, etc., after grain is removed. With portable bulkheads to confine grain to specific areas in the building, it can be used for two purposes at once.

Capacity

To estimate the capacity of a level filled flat storage building use Eq. 4-2.

Eq 4-2.

$$LFC = L \times W \times D \times 0.8$$

LFC = level filled storage capacity, bu
L = storage length, ft
W = storage width, ft
D = grain depth at sidewall, ft
0.8 = conversion factor, ft^3 to bu, bu/ft^3

For example, a 100′ long by 40′ wide flat storage filled to a level depth of 6′ holds an estimated 19,200 bu (100′ × 40′ × 6′ × 0.8 bu/ft^3).

To store more grain in a flat storage building, grain can be peaked. Although this allows for more storage capacity, it is more difficult to properly check and monitor grain in storage. It is considerably more difficult to walk on peaked grain than level grain for checking.

Most dry grains peak at an angle of 18°-28°, Table 4-1. The actual angle depends on grain type, moisture content, the amount of fines and foreign material in the grain, and height of drop. Minimum and maximum values are shown in the table to illustrate the range of filling angles that can occur in peaked grain storage. The actual peak height and storage capacity depends on the filling angle. The peak height can be estimated from Eq 4-3. For estimating grain capacity, use the average slope factor. The maximum filling angle should be used when evaluating the structural strength of a building used for grain storage.

Eq 4-3.

$$\text{Peak} = (\tfrac{1}{2} \times W \times SF) + D$$

Peak = height of peaked grain, ft
W = storage width, ft
SF = slope factor, Table 4-1
D = grain depth at sidewall, ft

For example, the peaked grain height in a 40′ wide building filled with soybeans or hard red spring wheat to a depth of 6′ at the sidewall can be calculated with Eq 4-3. From Table 4-1 the average filling angle is 25° and the slope factor is 0.47. The calculated peaked grain height is 15.4′ (½ × 40′ × 0.47 + 6′).

Table 4-1. Filling angles and corresponding slope factors.
Filling angles are affected by grain moisture content, foreign material, and loading method and can vary from these values. When estimating storage capacity, use the **average** angle.

	Average		Minimum		Maximum	
Crop	angle,°	Slope factor	angle,°	Slope factor	angle,°	Slope factor
Barley	28	0.53	24	0.45	34	0.67
Corn	23	0.42	21	0.38	26	0.49
Oats	28	0.53	24	0.45	32	0.62
Grain sorghum	29	0.55	27	0.51	33	0.65
Soybeans	25	0.47	22	0.40	29	0.55
Sunflower						
non-oil	28	0.53	20	0.36	40	0.84
oil	27	0.51	18	0.32	32	0.62
Durum wheat	23	0.42	22	0.40	25	0.47
Hard red spring wheat	25	0.47	19	0.34	38	0.78

The grain volume in a typical peaked flat storage building can be estimated by breaking it into three separate volumes, Fig 4-3. Use Eqs 4-4 to 4-7 to calculate the separate volumes. Multiply the total volume (V) by 0.8 bu/ft^3 to estimate the storage capacity.

Eq 4-4.

$$V1 = W \times L \times D$$

Eq 4-5.

$$V2 = \tfrac{1}{4} \times W \times W \times (L - W) \times SF$$

Eq 4-6.

$$V3 = \tfrac{1}{12} \times W \times W \times W \times SF$$

Eq 4-7.

$$V = V1 + V2 + (2 \times V3)$$

V = total volume, ft^3
V1 = level fill volume, ft^3
V2 = middle triangular peak volume, ft^3
V3 = peak volume at one end, ft^3
W = building width, ft
L = building length, ft
D = grain depth at sidewall, ft
SF = slope factor, Table 4-1

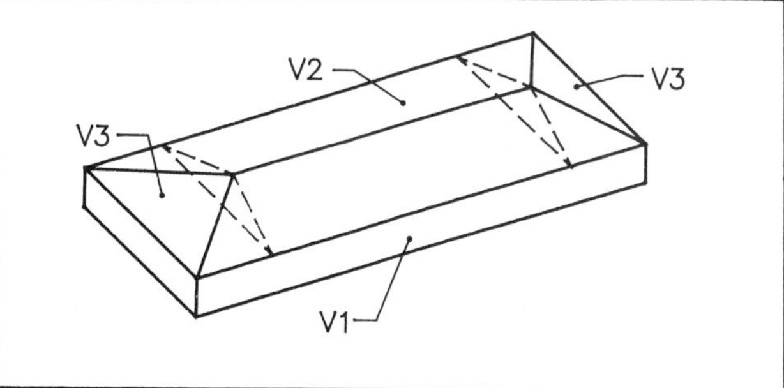

Fig 4-3. The three grain volumes of a flat storage.

Example 4-1:

Calculate the volume of a 40′ wide by 100′ long flat grain storage building with corn 6′ deep at the sidewalls and peaked at an average filling angle of 23°. From Table 4-1, the slope factor is 0.42.

Solution:

V1 = 40′ × 100′ × 6′ = 24,000 ft^3

V2 = ¼ × 40′ × 40′ × (100′ − 40′) × 0.42
= 10,080 ft^3

V3 = 1/12 × 40′ × 40′ × 40′ × 0.42
= 2,240 ft^3

Total volume = 24,000 + 10,080 + (2 × 2,240) = 38,560 ft^3

Estimated storage capacity = 38,560 ft^3 × 0.8 bu/ft^3 = 30,848 bu

Estimated peak height = (½ × 40′ × 0.42) + 6′ = 14.4′

If grain is stored against the walls, they must be adequately reinforced to resist the large outward grain pressure. Structural failures not only cause building and grain losses, but can also cause death.

Managing grain

Flat storages are not easily adaptable or economical for grain drying. It is difficult to uniformly load a flat storage to achieve a level grain surface required for even drying. Optimum drying air distribution is also difficult. Perforated floors provide the best airflow, but are expensive and difficult to install and maintain in a flat storage.

Dry grain aeration in flat storages generally uses ducts on the floor for air distribution. Mechanization of grain handling in flat storage is usually more difficult than in round grain bins. Grain is usually loaded into a flat storage with an overhead conveyor the length of the storage or by moving a portable inclined conveyor as the storage fills. Unloading is usually with portable augers, conveyors in the floor, a tractor with a bucket, a skid-steer loader, or a pneumatic conveyor. Unloading usually must work around surface aeration ducts, which are removed and stacked as unloading progresses. Some structures have diagonal wall ties to the floor, which also interferes with grain unloading. Another unloading

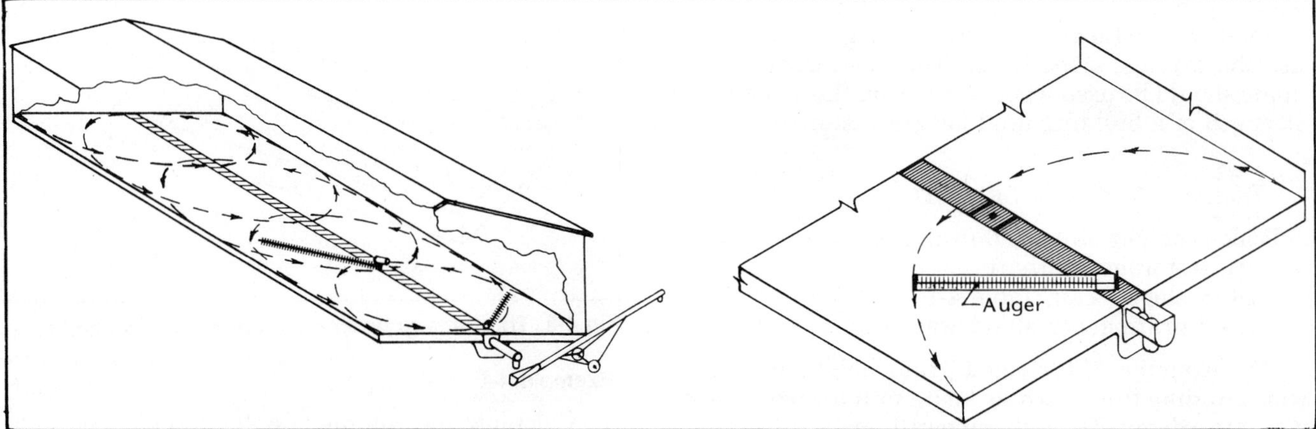

Fig 4-4. Unloading flat storages with augers.

option is a central conveyor recessed into the floor with a sweep auger, Fig 4-4. This system is convenient, but costly.

Install a vapor barrier to keep ground moisture from moving up through the floor into the grain. The floor of a flat storage should be at least 6″ above grade to minimize water problems. Drainage away from the storage is needed to prevent moisture buildup on the floor and around the storage.

Rodents and birds can be difficult to control in flat storages. The grain can be contaminated from mud or manure left in the storage by vehicles. Grease and hydraulic fluid on the floor can also contaminate grain. Thoroughly clean the building before storing grain, especially if fertilizer was stored in the building. Do not store pesticides, which can be toxic, in buildings that will store grain.

Consider these guidelines for successful flat grain storage:

- Keep it simple—do not try to dry in the structure; use surface aeration ducts; design to use portable equipment for loading and unloading, working inside the building if possible.
- Consider a combination of round bins and flat storages. Loading the flat storage last and unloading it first maximizes its alternate uses. In some years, all or part of the building may not be used for grain storage.
- Do not disregard permanent grain loading and unloading equipment when planning a flat grain storage. Future expansion and improvement may include the flat storage as a permanent storage.

Location

Do not let the dual use of a flat storage foul up the facility location such that it never gives good performance for either use. One use must dominate the location decision. If the primary purpose is grain storage, the importance of a good grain handling system should be the primary consideration. If the building is used for occasional emergency grain storage, locate the building for maximum convenience for its most commonly used function.

Constructing a machinery building as a temporary grain storage is sometimes advantageous. The returns from grain storage, such as government storage programs, can sometimes pay part or all of the building cost in 2 to 3 years, freeing the investment to build permanent grain storage. In this case, design and locate the building primarily for its ultimate use with grain storage considerations secondary. Movable bulkheads or bins without roofs can be placed in the building so sidewalls do not need to be designed for grain loads.

Existing Buildings

Many existing buildings or structures can be used to meet storage needs. Grain can be stored in some existing or temporary structures with relatively minor remodeling. However, converting a machine shed, cattle barn, etc., for deep grain storage requires major remodeling.

Examples of other buildings that can be remodeled for grain storage include corn cribs, empty livestock buildings, machine sheds, barns, upright silos, and horizontal silos. Some temporary storage ideas include reinforced plastic and steel mesh bins, hay bales lined with plastic, portable bulkheads, round plastic silage tubes, and piles. All of these can be successfully used provided proper grain management practices, including grain aeration, are implemented. It is absolutely essential that all existing buildings and temporary structures be adequately reinforced to withstand the expected grain pressures to prevent structural failures.

Converted corn cribs

Ear corn cribs can be converted to grain storage. Get Midwest Plan Service publication AED-12, *Remodeling Corn Cribs for Small Grain Storage* for procedures to reinforce corn cribs and make them grain- and weather-tight for shelled corn or small grain.

Wide converted ear corn cribs may have to be aerated. Cribs 10′ wide or less usually do not require aeration. However, as crib width increases, the grain in the center is insulated by surrounding grain which keeps it warmer than the outside grain. The higher temperature differences between the center and outside grain increases convective air movement, which can lead to moisture migration and condensation.

Taller cribs tend to increase convection currents even more because of the chimney effect. An aeration system is essential in wide, tall cribs. If a crib has recessed shelling trenches, they can be used for aeration ducts. When the air ducts are at or in the floor, the sidewalls must be tight to prevent air leakage.

Converted corn cribs are usually not satisfactory for grain drying. It is difficult to uniformly fill and maintain a level grain surface which is essential for grain drying.

Many converted corn cribs have been incorporated into a convenient grain handling system. Often the location of converted corn cribs is best suited for emergency storage or for grain that does not need to be dried, such as soybeans.

Remodeling considerations

Grain in a building pushes out against the walls, so the building frame and lining material must withstand large horizontal loads. Fastenings, anchorage, and the foundation must withstand these loads without breaking or excessive movement. Before storing grain in an existing building, determine the strength of the walls, floor, and other structural members to be sure they can withstand the expected pressures. Structural failures not only cause serious structural and grain losses, but can also cause death. Consult your state extension agricultural engineer or someone experienced in grain storage to structurally evaluate existing buildings for grain storage.

Shallow grain piles are easier to contain than deep ones. For example, the total horizontal load from a 10′ deep pile is 6.25 times greater than the load from a 4′ pile. The horizontal pressure exerted by a peaked pile of grain is greater than the load from a level filled pile of grain even though the grain depths are the same at the wall. The load from a peaked pile of grain is about ⅓ greater a level pile of grain (multiply by 1.33).

Lay a plastic vapor barrier between the floor and grain to prevent moisture from migrating through the floor into the grain. If a vapor barrier was installed under the floor, a second vapor barrier is not needed. Check the roofing for watertightness. Reduce bird and rodent damage by screening any necessary openings. Minimize ledges and cracks that can hold or trap old grain, because they are ideal for insect breeding and harborage. Maintain a rodent bait program. Consult your extension entomologist about insecticide treatment.

Upright silos for dry grain storage

Upright, roofed silos of any type can store dry grain. This includes silos made from concrete and clay tile with external steel hoops, cast concrete, metal, fiberglass, and plastic.

Take the following precautions when using upright silos for dry grain:

- Be sure the silo is structurally sound. If it was designed for high moisture grain, it should be adequate for dry grain. Check with the structure's original manufacturer or your state extension agricultural engineer for recommended structural reinforcing. Avoid old silos, especially if reinforcing steel is imbedded in the wall. Older silos with weathered walls may not be able to withstand the pressure exerted by the grain and collapse. Walls must be sound so moisture does not enter.
- In concrete stave silos, the maximum dry grain storage depth is about 50′. Dry grain puts a higher vertical load on the walls which could cause the staves to break. Before storing grain deeper than 50′, consult your state agricultural engineer to determine the structural adequacy of the silo.
- Aerate the silo to prevent convection currents and moisture migration. Use positive pressure aeration for silos adapted for dry grain. Fig 4-5 shows a suggested aeration system for oxygen-limiting silos. Fig 4-6 shows an aeration system for conventional upright silos.
- Load at the center of the silo to prevent off-center loading which could cause structural damage.
- Unload the silo from the bottom center—do not draw from the side because unbalanced loads can cause structural damage or failure. Use a portable auger to unload grain through the aeration duct.
- Be sure silo roof and stave joints do not leak during blowing rains.
- Some type of floor and vapor barrier must be installed. This can be a concrete floor or a layer of sand at least 6″ deep with a plastic vapor barrier. If you expect to use the upright silo for high moisture grain or silage in the future, a sand floor allows more flexibility.

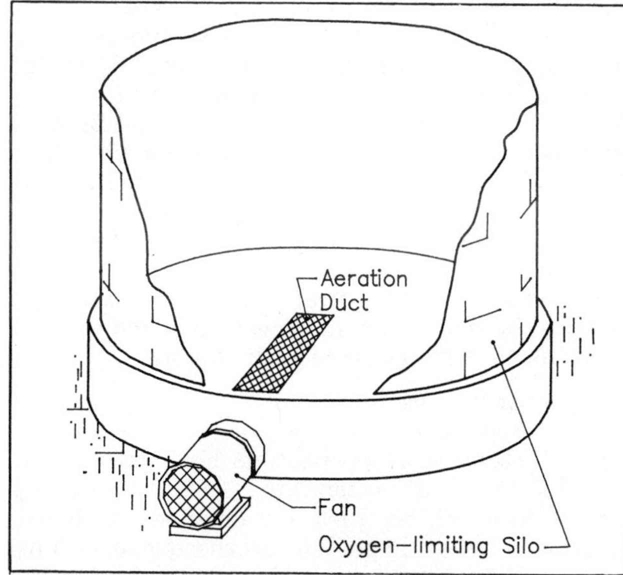

Fig 4-5. Aeration duct in an oxygen-limiting silo.

High moisture grain storage

High moisture grain storage is an excellent method for storing grain to be fed. High moisture grains include all feed grains and high moisture

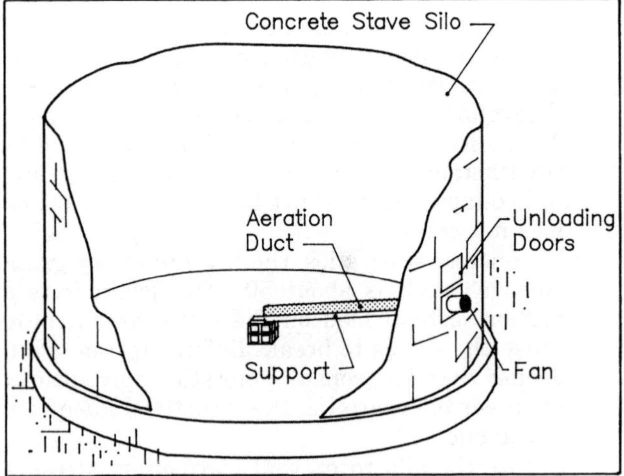

Fig 4-6. Aeration duct in a concrete stave silo.
Unload through the aeration duct with a portable auger.

ground ear corn. High moisture grain can be stored in bottom unloading, sealed structures that are poured concrete, concrete stave, or metal.

Top unloaders work well on either whole or ground grain in conventional upright silos. Bottom unloaders do not operate well with cracked or ground grain without special unloading equipment, because the grain bridges over unloading augers. Bottom unloading high moisture grain works only with sealed, oxygen-limiting structures.

High moisture grain can also be successfully stored in horizontal silos.

Consider locating high moisture grain storage facilities near the grain center to use the same grain conveying and processing equipment for both wet and dry grain.

Chemical preservatives are also used for storing high moisture grain. Chemically preserved grain can only be marketed for livestock feed. Chemicals are corrosive to concrete and metal; **use special coatings to protect interior wall surfaces.** Apply the coating before any chemically preserved grain is stored; otherwise surface preparation will be difficult. Wear personal protective equipment when handling chemical preservatives and working around freshly treated grain. Chemically preserved grain requires proper aeration to prevent convection currents and resulting moisture migration.

Most state agricultural engineering extension services have more specific information and publications about high moisture grain storage.

Hopper-bottom bins

Hopper-bottom bins allow gravity unloading and are self-cleaning. Hopper-bottom bins are more expensive than flat-bottom bins, so use them only where they will be filled and emptied frequently, such as in wet grain holding, overhead loadout bins, wet holding above a high temperature automatic batch dryer, or feed processing. An elevated hopper-bottom bin requires additional support which makes it even more expensive. Most hopper-bottom storages are commercially available round steel units. However, they can also be square steel bins or built with lumber or concrete.

Piles

Piles can be successfully used for emergency storage (1 to 3 mo). Grain losses are higher than in permanent storages. Proper grain management is essential to minimize losses. Grain should be dry and reasonably clean. Pockets of wet grain or trash can cause spoilage. Aeration and grain checking techniques are especially important. Contact your state extension agricultural engineer for additional details.

Remodeling Post-Frame Buildings

In this book, a post-frame building can be either a square post or round pole building. These remodeling ideas can be used for either.

To store grain in a post-frame building, extra posts may be needed, depending on grain depth and the building's design. Similar major changes may be needed in wood stud or steel frame buildings. Consult an engineer experienced in grain storage design for recommended techniques to properly reinforce walls of existing structures.

Some remodeling ideas are presented in this book for storing grain up to 4′ deep in post-frame buildings. Follow construction details exactly to eliminate the chance of a building failure. The following ideas are for ground level storages only. Before adding grain storage to a mow floor or other elevated floor, consult an engineer experienced in grain storage design.

Contamination from wood preservatives may become a problem in post-frame buildings. Federal regulations restrict use of preservatives in contact with grain and feed. Do not use creosote or pentachlorophenal where feed or grain is stored.

4′ High plywood liner

Lining the lower 4′ of existing post-frame building walls is fairly economical and does not interfere with future uses of the building. A paved floor makes unloading easier.

With at least 6″x6″ posts, 8′ o.c., 4′ in the ground, and no more than 14′ above the ground, grain can be safely added up to 4′ deep at the sidewalls. For building frames with posts smaller than 6″x6″, consult your state extension agricultural engineer.

Fig 4-7 shows construction details for adding a 4′ high wall between the posts to contain the grain. Cut a 2x4 sill and plate to fit between the posts. Assemble the frame with 16d nails. Attach the 2x6 sill and 2x6 plate to the 2x4 studs with two 16d nails at each end. Tip the frame into place. Fasten the sills and plates to the posts with framing anchors. Line the wall with 4′x8′ 5/8″ or 3/4″ plywood sheets. Select plywood with an Identification Index of 42/20 Ext. Or, use 5/8″ CDX plywood if from Group 1 or 2 species and 3/4″ plywood if from Group 3, 4, or 5 species.

To keep grain from falling between the grain wall and the building wall, add a baffle from the top of the grain wall to a wall girt. Do not pile grain deeper than 4′ against the building wall.

An alternative to the construction in Fig 4-7 is to move the new framing inside the posts. The sill and plate can be continuous across one or more posts; add

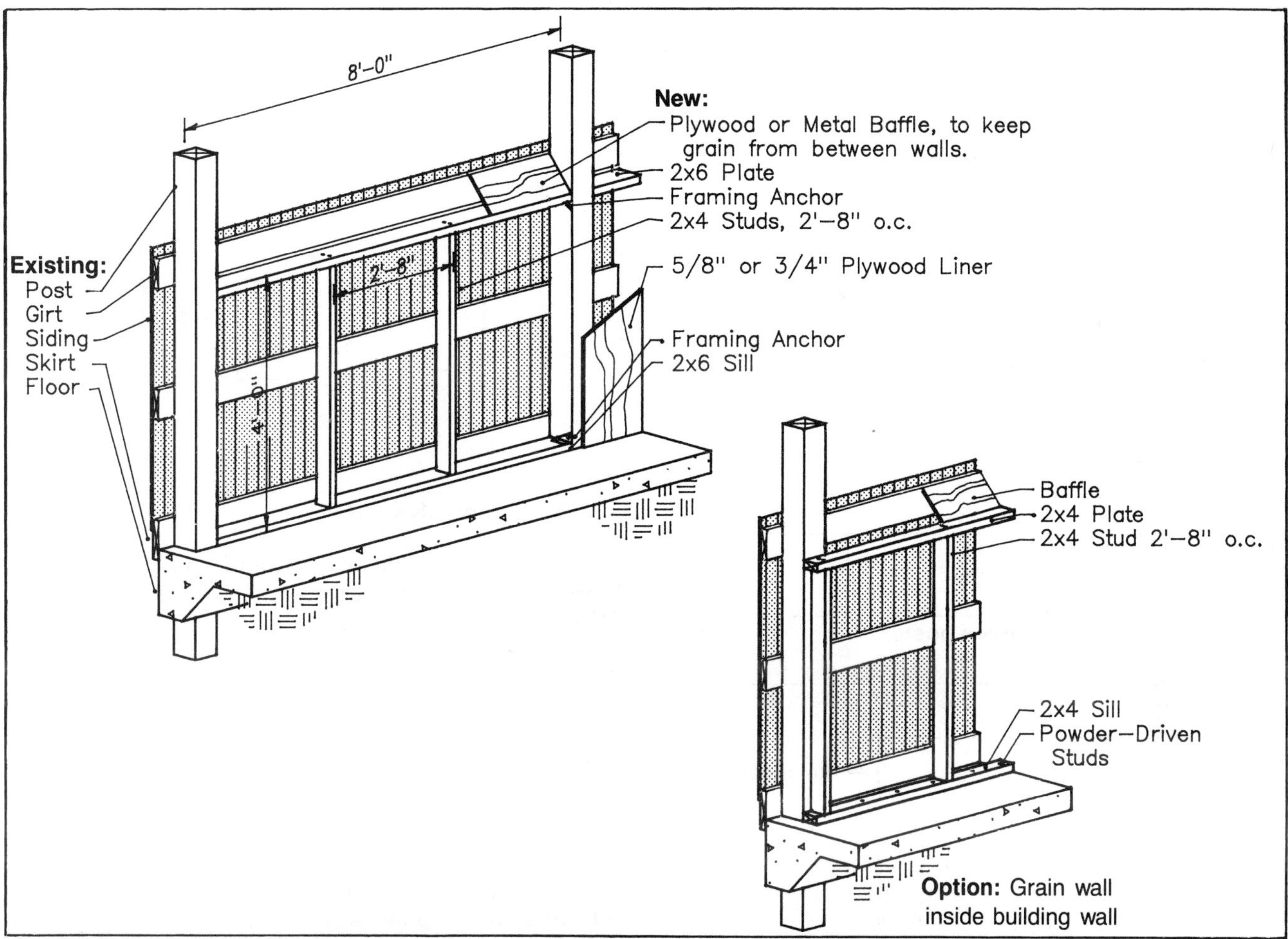

Fig 4-7. Grain liner for post-frame buildings.
Fasten plate and sill to 2x4 studs before installing framing in the wall. Drive two 16d nails through sill and plate into ends of each stud.
Studs are flush with inside face of posts.
Lumber is No. 2 or better, Southern Pine or equivalent.

a stud at each post to support the plywood. Anchor the sill to the floor with angle irons and anchor bolts or with anchor nails driven through the sill with a stud gun.

4′ High concrete wall

A 4′ concrete wall can be cast between posts of new or existing post-frame buildings to reinforce it for grain storage, Fig 4-8. Concrete walls can be easily cast between posts up to 8′ o.c. When building a new building tie the wall and floor together with reinforcing steel. In new construction, cast the wall at least 24″ below grade. To add a concrete wall to existing buildings with floors, cast the wall between posts on top of the floor.

For specific design and construction techniques get AED-28, *Cast-In-Place Concrete Walls For Farm and Home*. For detailed information about the concrete quality required for farm buildings and other information about concrete placement, see *Farm and Home Concrete,* AED-26. These publications are available from the extension agricultural engineer at any of the institutions on the inside front cover or Midwest Plan Service.

Bin rings

Round bin rings can be conveniently and economically erected inside an existing building to provide emergency grain storage. The rings can be either commercially available steel bin rings or round plywood bins.

Two or three steel rings from round grain bins can be set on an existing floor. Install an unloading auger to the center of the bin rings for emptying. Anchor the bin rings as recommended by the manufacturer.

The round shape may leave building corners that are difficult to utilize. But rings can hold more grain with less remodeling. For example, a 40′ square pile which is 4′ deep at the walls and peaked at 23° holds 8,742 bu. Round bin rings, 40′ diameter and 7′-6″ high in the same building with grain peaked at 23°, hold 10,357 bu.

Round Plywood Bins

Plywood bins, 4′ or 8′ deep and 10′-20′ in diameter, can be made by nailing plywood sheets and 2x4 nailers into a long strip and bending to form a circle, Fig 4-10. Fasteners and nails are critical in keeping the bin from bursting.

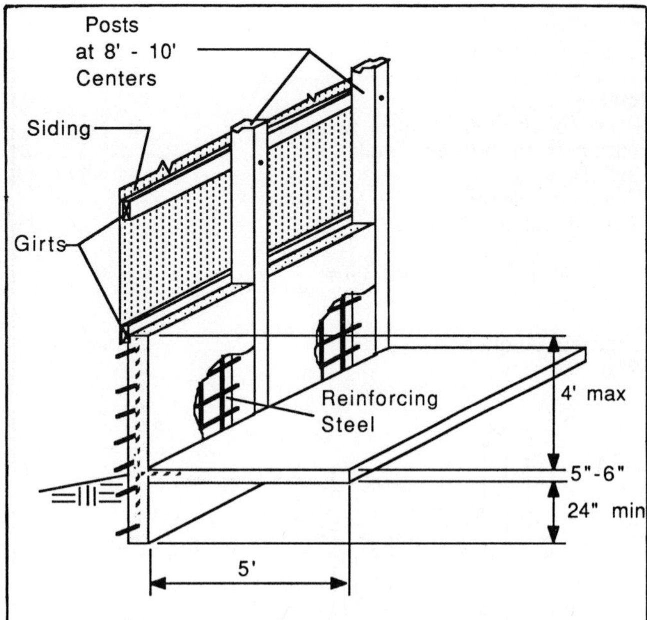

Fig 4-8. Concrete wall between posts.

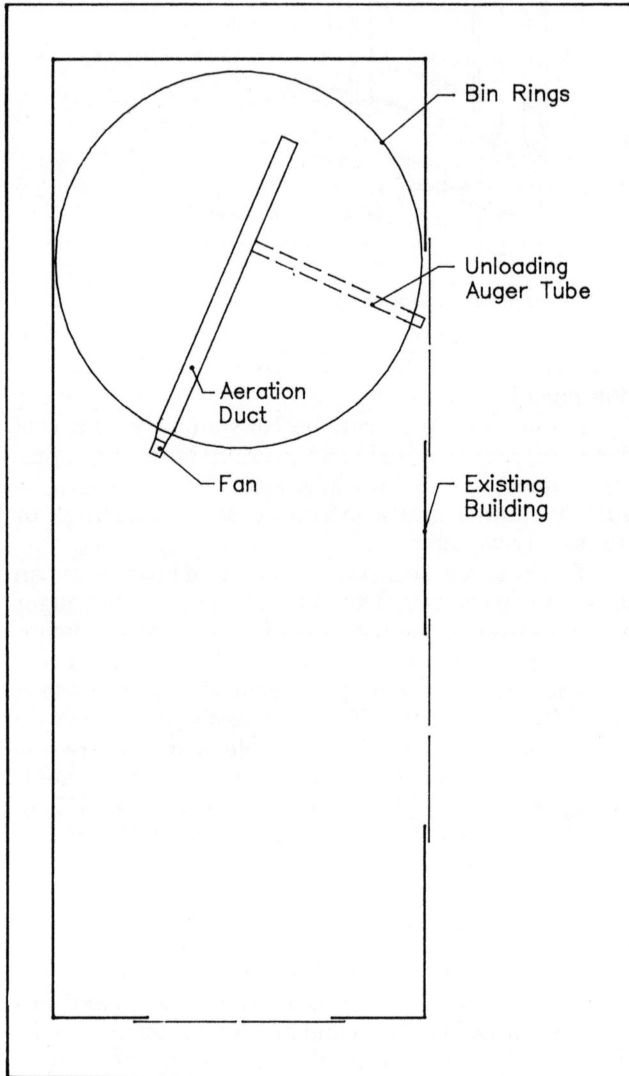

Fig 4-9. Bin rings added to an existing building.

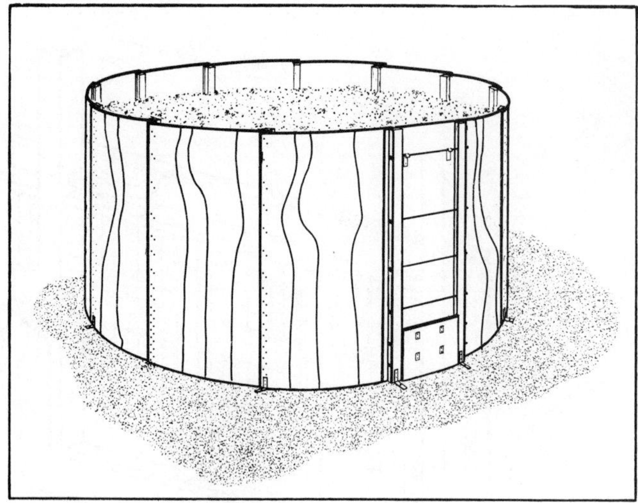

Fig 4-10. Round plywood bin.

These bin designs are available from the Canada Plan Service, Agriculture Canada, Research Branch ESRI, Ottawa, Canada, K1A OC6. The plans are plan number 8421, Circular Portable Fertilizer Bin—8′ High, and plan number 8422, Circular Portable Fertilizer Bin—4′ High.

Movable Grain Walls

Plan mwps-73210, available from Midwest Plan Service, details 6′, 8′, 10′ and 12′ deep movable grain storage walls. They are built of plywood and lumber, are self-supporting, and can convert all or part of a building to grain storage.

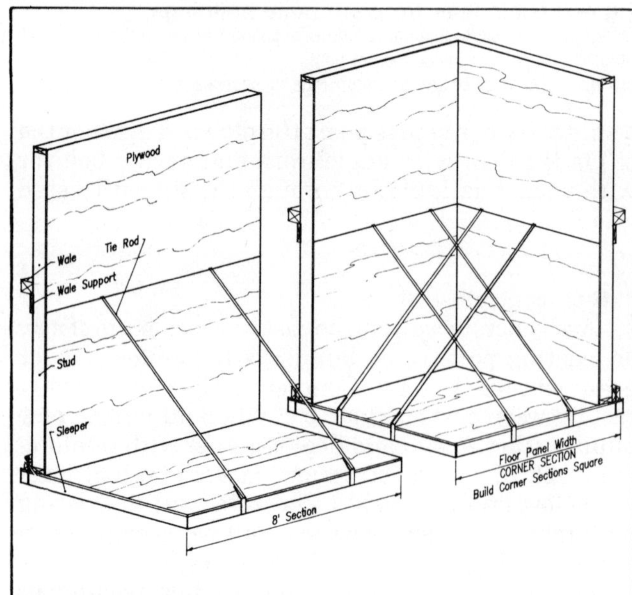

Fig 4-11. Movable grain walls.

Managing Grain in Storage

Properly managing grain in storage is important to maintain grain quality after it is harvested. Putting grain in storage does not improve grain quality. Generally, to maintain quality, all biological and in-

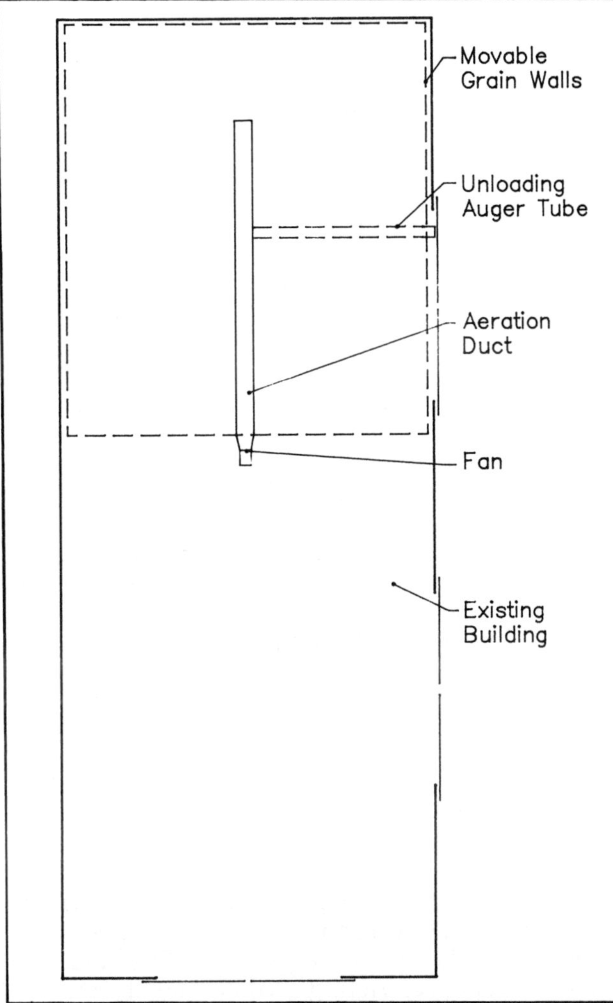

Fig 4-12. Movable grain walls in an existing building.

sect activity must be controlled. Factors that can cause grain to go out of condition are:

- Initial grain quality.
- Grain moisture content.
- Grain temperature.
- Amount and distribution of fines and foreign material.
- Presence of insects.

Moisture content

The average moisture content of all grain in a storage does not determine crop storability. Spoilage can occur in isolated locations or pockets where grain moisture is high. Table 4-2 gives maximum recommended moisture content for several grains.

Grain temperature

Grain temperature must be controlled to minimize moisture migration which is a major storage problem. **More stored dry grain goes out of condition because temperatures are not controlled than for any other reason.** Lower grain temperatures decrease biological (mostly molds) and insect activity and increase safe storage periods.

Grain is a good insulator, so it does not cool uniformly as outside temperatures drop during late fall and winter. Air near the bin wall cools and settles

Table 4-2. Maximum moisture contents for safe grain storage.
Values for good quality, clean grain and aerated storage. Reduce 1% for poor quality grain, such as grain damaged by blight, draught, etc.

Grain type & storage time	Maximum moisture content for safe storage %
Shelled corn and sorghum	
Sold as #2 grain by spring	15½
Stored 6-12 mo	14
Stored more than 1 yr	13
Soybeans	
Sold by spring	14
Stored up to 1 yr	12
Stored more than 1 yr	11
Wheat, oats, barley	
Stored up to 6 mo	14
Stored more than 6 mo	13
Sunflower	
Stored up to 6 mo	10
Stored more than 6 mo	8
Flaxseed	
Stored up to 6 mo	9
Stored more than 6 mo	7
Edible beans	
Stored up to 6 mo	16
Stored more than 6 mo	14

toward the bin bottom creating a convection current. The air then rises up through the warm grain picking up moisture in the form of water vapor. The air continues to move toward cooler grain near the grain surface, where the moisture condenses and can cause spoilage, Fig 4-13.

The most common location of wet or spoiled grain is at the center top of the bin. Another location for storage problem symptoms is the grain near the bin wall, often the cold north wall.

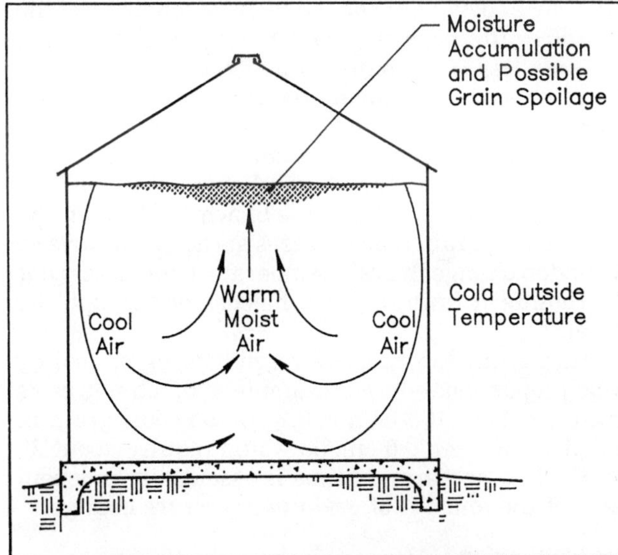

Fig 4-13. Typical moisture migration pattern.
Moisture migration usually occurs during fall and winter. Increased moisture content and crusting usually occur at the top center of the bin (the dark area). Moisture may also accumulate in cold grain near the bin wall, often near the cold north wall.

Aeration Systems

Aeration is essential for successful dry grain storage. A properly operated aeration system moves air through grain to control grain temperature, reduce biological and insect activity, and prevent moisture migration. Aerate grain during fall to uniformly cool the grain to below 40 F to prevent moisture migration. In the northern parts of the Midwest (South Dakota, North Dakota, Minnesota, and Wisconsin), cool stored grain to below 35 F because of colder average winter temperatures. During the spring uniformly warm grain to between 50 to 60 F for optimum storage during spring and summer.

Proper operating and management procedures must be followed with any aeration system. Even the best designed aeration system can fail if not operated correctly. A modestly designed system and proper management is better than an elaborate system and little management. For more about proper operation of grain aeration systems, get Midwest Plan Service publication AED-20, *Managing Dry Grain in Storage,* from your county extension office or state extension agricultural engineer.

Airflow rates

Grain can be successfully aerated with airflow rates as low as 0.05 (1⁄20) cfm/bu to over 1 cfm/bu, which is a common rate for drying grain. Although drying fans can be used for aeration, low capacity aeration fans cannot be used for drying because they do not move enough air to dry the grain before spoilage occurs.

Airflow rate determines the time required to complete one aeration cycle. An aeration cycle is the time it takes to change the temperature of all grain in the bin or for the cooling front in the fall and the warming front in the spring to pass through the bin. For example, at the common aeration airflow rate of 0.1 cfm/bu, it requires about 150 hr to change grain temperature. This varies depending on the relative condition of the air and grain. Typically, warming grain in the spring requires less time than cooling grain in the fall.

To be sure of complete cooling or warming, you must monitor the change in grain temperature. On a negative pressure or suction aeration system, check the temperature at the fan discharge. With positive pressure systems, where air is blown up through the grain, check grain temperatures at the grain surface. A sudden change in grain temperature indicates that the cooling or warming front has passed through the grain.

Higher airflow rates decrease heating or cooling time proportionately. For example, if 0.1 cfm/bu cools grain in about 150 hr and 0.2 cfm/bu cools grain in about 75 hr, then 0.5 cfm/bu would require about 30 hr. Cooling or warming time is essentially independent of the amount of grain temperature change.

Air distribution

Proper design of an aeration system and arrangement is important for proper air distribution. Size the duct cross-sectional area so the duct air velocity is less than 1,500 ft/min (fpm). High duct air velocities increase fan power requirements and can cause non-uniform airflow from the duct. Do not exceed 1,500 fpm without proper engineering advice. To achieve 1,500 fpm air velocity, provide 1 ft^2 of area for each 1,500 cfm.

There must also be enough perforated area to allow air to exit without excessive restriction. When using standard perforated metal duct or flooring provide 1 ft^2 of perforated surface area per 30 cfm.

Air must have an unrestricted path into and out of the storage. Provide at least 1 ft^2 of opening in the roof, gable end, or eaves for each 1,000 cfm of airflow.

In round bins, full perforated floors are used with high airflow systems to provide more flexibility for different uses, but are expensive for aeration systems. Therefore, many storage bins are equipped with aeration pads or ducts for air distribution. Sub-floor ducts work well in deep round bins and do not hamper grain removal. Fig 4-14 shows common sub-floor duct arrangements for round bins. Flat storages and piles almost always use aeration ducts.

Install above-floor ducts in flat storages for easy removal during grain unloading. Ducts are usually placed directly on the floor and are not fastened down. Carefully fill the storage by directing grain around the duct to prevent them from shifting out of place. In storages with sub-floor unloading equipment and sweep augers, use sub-floor aeration ducts, Fig 4-15.

Good judgement is required for duct spacing and arrangement to provide uniform air distribution. With peaked grain storage, if grain is delivered to the center of the building and allowed to flow to the walls, the grain near the center contains more fines and will be packed more, restricting airflow near the center. In a building up to 40′ wide, a single properly sized duct may effectively distribute air, Fig 4-15a. Buildings between 40′ to 60′ wide with peaked grain require at least two properly sized ducts. Limit duct length to 100′ per fan. For example, in a 160′ building use two 80′ ducts with a fan on each duct.

In level filled storages, for effective air distribution make the duct spacing equal to the grain depth. Place the ducts nearest the wall only ½ the grain depth from the wall, Fig 4-16.

In this book, general design information for aeration systems has been included. For more specific information about aeration system selection, sizing, and layout, get AED-29, *Aeration Systems For Grain Storage,* and contact your state extension agricultural engineer or a knowledgeable equipment dealer.

Grain Loading

Broken grain and foreign material, or fines, can create two problems in stored grain, particularly when they accumulate in pockets. First, broken kernels are more susceptible to spoilage than unbroken ones. Second, airflow from aeration fans tends to go around pockets of fines so they cool more slowly. These pockets often develop into hot spots that result in spoiled grain.

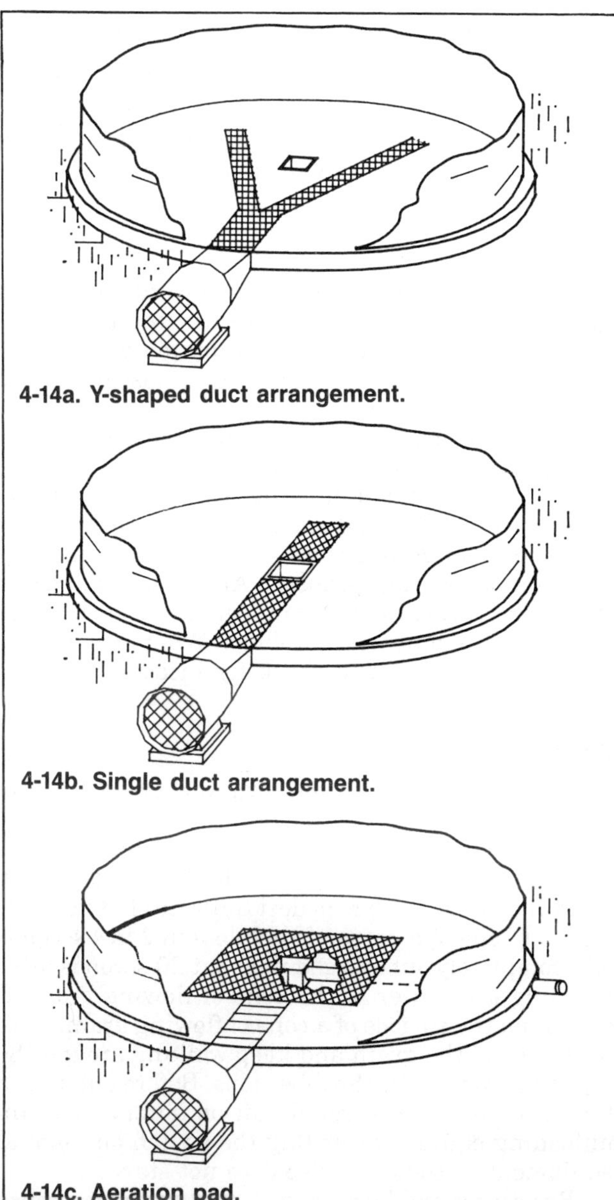

4-14a. Y-shaped duct arrangement.

4-14b. Single duct arrangement.

4-14c. Aeration pad.

Fig 4-14. Common subfloor aeration duct patterns for round bins.
Duct arrangement may be affected by grain depth. Make the perforated floor sections easily removable for cleaning the duct.

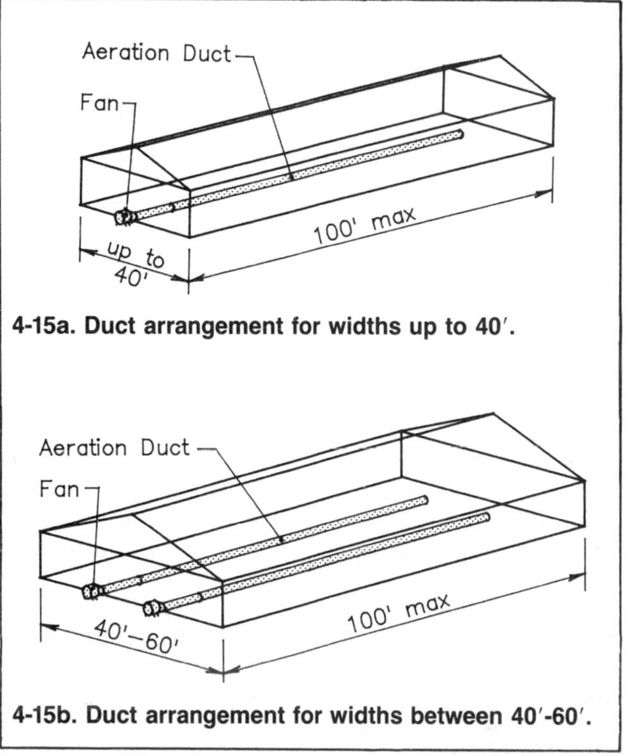

4-15a. Duct arrangement for widths up to 40′.

4-15b. Duct arrangement for widths between 40′-60′.

Fig 4-15. Typical duct arrangements for some large flat storages.
Successful flat storage aeration depends on proper sizing and location of the ducts. Limit duct length to 100′ per fan. For buildings up to 200′ long, use two ducts with a fan on each duct.

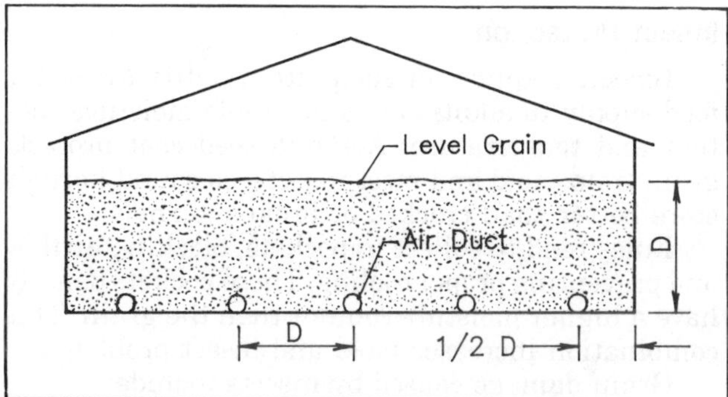

Fig 4-16. Duct arrangement in flat storage with level grain.

Strive to minimize the fines produced from harvesting, drying, and handling, rather than later trying to solve resulting storage problems. Consider two grain storage management techniques that avoid problems from fines that do go into the storage:

- The most common technique is to use a grain spreader to minimize concentration of fines. If a power spreader is not used, install a grain cone in the bin to break up the inflow of grain and partially spread the fines.
- Grain can also be cleaned before putting it in a bin to improve storability. Unfortunately, cleaned grain may have no greater market value. Fines do have value, both as livestock feed and as added weight in market grain. It makes sense for livestock farmers to clean all their feed grain because they can feed the fines. Unless fines are causing serious storage problems or grade reduction, cash grain farmers may lose money by cleaning grain unless they can sell the gleanings to a livestock farmer or grain elevator.

When planning a new grain storage and drying system, plan for a way to include grain cleaning now or in the future. The grain grading rules may change to make on-farm grain cleaning more profitable.

Grain Peaking

Most dry grain peaks at an angle of 18°-28° for center filling without a distributor. The volume of grain that can be stored in the peak of a round bin can be estimated with Eq 4-8. Although it is tempting to store those extra bushels, peaked grain is more difficult to uniformly aerate which can cause moisture migration problems.

Eq 4-8.

$$PC = 0.131 \times D \times D \times D \times SF \times 0.8$$

PC = estimated capacity of the peak of a round bin, bu
0.131 = constant
D = bin diameter, ft
SF = slope factor, Table 4-1
0.8 = conversion factor, ft^3 to bu, bu/ft^3

For example, the estimated capacity of the peak in a 24-foot diameter round bin of corn with an average filling angle of 23° is 608 bushels (0.131 × 24 × 24 × 24 × 0.42 × 0.8).

Peaking also makes it difficult and dangerous to enter the bin for checking grain. Dust and high temperatures during the summer makes it dangerous to enter the small space between the roof and peaked grain. Entry into a peaked round bin storage should be prohibited during the summer. Shifting grain may block the exit. If grain is peaked in a round bin during harvest, withdraw the peak immediately after harvest.

Grain Unloading

For convenient grain bin unloading, use unloading and sweep augers. See Chapter 2 for more information about different types of grain unloading equipment.

Insect Protection

Insects require an adequate, readily available food supply in addition to a favorable moisture content and temperature. A sound seed coat protects grain from mold and most insects, so sound kernels store better than cracked or broken kernels.

Fines tend to collect in pockets, blocking airflow and preventing proper cooling. These pockets usually have a higher moisture content than the grain. This combination increases mold and insect problems.

Grain damage caused by insects include:

- Bored holes and the disappearance of much of the inside of the kernels.
- Injury to the grain germs.
- Heating and consequent condensation and grain molding.
- Contamination with insect bodies, body parts, excrement, and webbing.

Limit insect damage by controlling grain temperature and moisture content. Insect activity decreases sharply at temperatures below about 60 F. Insects are nearly stopped below 50 F. Fumigation is not recommended at grain temperatures below 60 F, because at low temperatures fumigants penetrate the seed coat more slowly.

To limit or prevent insect development:

- Thoroughly clean and spray all storage surfaces with an approved residual insecticide at least two weeks before grain is stored.
- Store grain in weathertight structures.
- Treat grain with an approved insecticide.
- Cool grain with proper aeration.
- Inspect grain at frequent and regular intervals.
- Clean fines from aeration ducts and under perforated floors or fumigate with an approved product. Fines provide an excellent breeding place for insects.

Checking

Check stored grain weekly during critical fall and spring months when outside air temperatures change rapidly and during summer when temperatures are high. Check grain at least every two weeks during winter.

Get into every bin and check for signs of moisture migration—wet, slimy grain, ice or frost accumulation, crusting, and/or heating. Check for odors by turning on the fan and smelling the air exhausting from the grain. A foul or musty odor is often an early indicator of a storage problem. Also check and record the temperature of the air. An increase in temperature also indicates a problem. When in a bin feel and probe the grain to check for insects and other problems.

Safety Practices

Review safety measures with workers and family members. Absolutely forbid entry into a bin or gravity unloaded vehicle when grain is flowing. Suffocation is a major cause of accidental death when handling grain. With a modest flowrate of 1,000 bu/hr from a 6″ auger, a person is helpless in 2 to 4 seconds and completely submerged in about 20 seconds after reaching the center of the cone of flowing grain. If caught on the surface of a cone of flowing grain, walk to the top of the grain and keep walking around the top of the cone until the flow stops. Before entering a bin, lock out the control circuit on automatic grain unloading equipment or flag the switch on manual equipment so someone else does not start it.

Breathing mold spores in stored grain can cause illness and can lead to chronic health problems. Wear a respirator capable of filtering fine dust when working in dusty or moldy grain. Maintain proper and effective shields and guards on hazardous equipment such as moving belts, chains, pulleys, gears, and shifts.

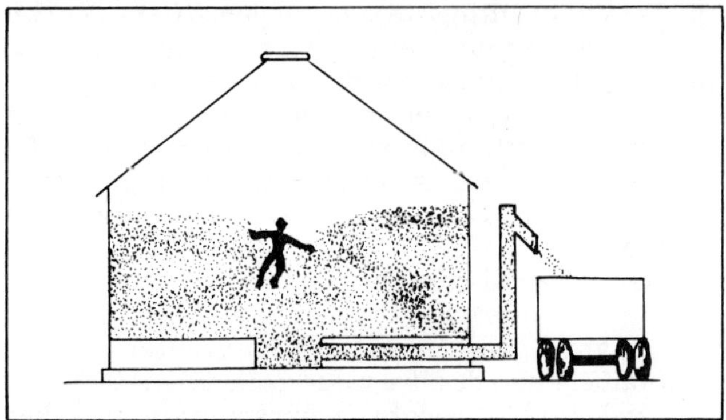

Fig 4-17. Flowing grain can trap you in seconds.

5. LOCATING AND DEVELOPING A GRAIN CENTER

Expansion or New

Carefully consider existing facilities when expanding or building a new system. A grain center built 15 years ago may have severe limitations: required capacity has at least doubled; handling equipment is worn and under-sized; dryers are the wrong type or size for current grain volumes; the closest bins are the smallest; and larger bins, which were added later, are farther away, Fig 5-1. Do not let existing facilities with marginal life and performance affect the location, layout, and performance of a new or expanded system for the next 25 years.

Consider building a new grain center at a different site or near existing facilities to take advantage of older bins for dry grain storage. If possible, move the best existing bins and equipment to the new site, to an existing feed processing facility, or to another farm—or sell them. Consider using the existing facility as a satellite grain center; or convert it to handle an alternate grain such as soybeans or wheat, to a feed processing center, or to long-term storage.

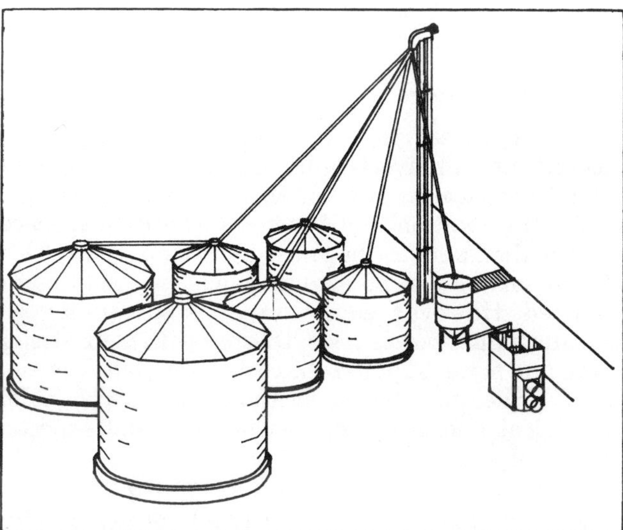

Fig 5-1. Scattered layout typical of old grain systems.
Smaller bins were located near the hub when the grain system was first built. As the operation grew, larger bins were added, but had to be located farther from the bucket elevator.

Central or Scattered Sites

Grain facilities are usually at one central site. A grain center works well for either a cash grain or grain-livestock farm. However, grain facilities are sometimes needed at more than one site. Consider your farm's needs when planning the grain center location.

Advantages of central facilities include:

- More efficient grain handling, drying, storage, and management.
- Push-button and automatic controls.
- Single electrical and fuel services for the entire system.
- Centrally located records and instruments.
- Dryers and other equipment do not have to be moved between sites.
- Security from theft and vandalism.
- Fewer all weather roads.

Advantages of separated facilities include:

- Smaller facilities are easier to divide if the farm business breaks up, partners divide the operation, or family status changes.
- Less fire risk than central facilities.
- Less competition with other enterprises for expansion space.
- Shorter transport distances and conveyors between storages and the facility center.
- Decreased traffic around the main farmstead.
- Flexibility to meet specific needs, such as storing landlord's grain separately.

More than one grain center may be needed if you operate on several farms and the landlords insist that their grain be stored on their land. If you have this situation and prefer a central grain center, consider:

- Marketing the landlord's grain at harvest time.
- Drying the landlord's grain at a central facility and storing it on his land. The number of trips may not increase because trucks bringing wet grain from a remote farm can haul dry grain on the return trip.
- Drying and storing grain centrally and renting storage space to the landlord.
- Designating a bin at the grain center for the landlord's grain. Establish grain ownership and bin cost sharing by contract in case the land or bin lease is terminated.
- Blending (co-mingle) all grain together and maintaining grain receipts and sales records like commercial elevators. You will need a scale and a contract on management responsibilities and ownership.
- Buying the landlord's grain at harvest or at a future date.

Site Selection

At each probable site, ask:

- Is there enough space for immediate needs?
- Does it have good drainage?
- Is it accessible to roads and grain or livestock production areas?
- What is the effect of system dust and noise?
- Is there room to expand?
- Is adequate electric power available or can it be economically extended to the site?

Space Requirements

Provide space for:

- Grain handling and cleaning equipment.
- Wet holding, drying, cooling, and storage bins.
- Dryer.
- Feed processing and storage.
- Electrical service and control.
- Management building for shelter, record keeping, and grain quality testing.
- Scale for weighing.
- Fuel storage.
- Roads.

All these facilities may not be needed now, but consider providing adequate space in the plan for future expansion and improvements. You may need to include: increased storage, a bucket elevator, pneumatic conveying, wet grain receiving, a new drying method, a switch to livestock or cash grain, a change in marketing plans, or feed processing.

Consider starting your plan with an area about 120'x120'. This should provide ample space for the original system and immediate expansion needs. Plan space for future expansion. For example, clear a nearby wooded area, remove a less useful building, or convert some pasture or cropland.

Drainage

Avoid areas with surface water runoff problems and high water tables. Select a site with surface drainage in all directions. If necessary, add earth fill to increase grade level. If surface drainage in all directions is not possible, divert water away from the grain center with waterways and terraces.

Regardless of the grain center's initial capacity or complexity, locate it and establish the grade based on future expansion. Move earth into an area **before** construction begins.

Location

Grain centers generate noise, dust, and traffic. Locate your center downwind of prevailing winds during the harvest season and at least 200' from houses, neighbors, and other areas occupied by people.

A location remote from the house reduces traffic nuisance and safety problems in the living area. Reduce risk of vandalism and theft with safeguards such as locked gates, locked unloading auger motors, and trespasser warning devices.

Provide convenient access to the main road. Consider the weight of grain transport vehicles when locating the grain center to avoid light-load bridges and culverts. Consider a by-pass road to take non-grain traffic such as employees, family, or visitors around the grain center.

Avoid access routes with drifting snow or ponding water. If possible, take advantage of winter sun to thaw snow and ice and increase worker comfort by locating roads and unloading augers on the south side of bins and buildings.

Grain center location relative to present or future livestock operations is determined by the feed delivery system—vehicle or conveyor. With vehicles (portable grinder-mixer, self-unloading wagon, truck), transport distance is not critical unless it is excessive. Considering time to start the vehicle, hitch the tongue, couple the PTO, open doors, gates, etc., it may make little difference whether you travel 500', 1,000', or a mile. Provide convenient, all-weather roads with a vehicle transport system. Conveyor delivery (auger, pneumatic, cable) over 500'-800' usually needs relay stations or intermediate power drives.

On large grain and livestock farms, consider separating feed processing from the grain center. Consider a satellite feed processing facility with one or two grain storage bins. Fill these bins at harvest and refill as needed with the same equipment and vehicles used at harvest and marketing.

Consider a grain center near the center of the total farming operation, especially if fields are separated by several miles. Minimizing the distance to all fields saves fuel and hauling time and can eliminate bottlenecks.

Locating a new grain center near a primary market can be convenient, but is usually not a very high priority. Grain is often not hauled to market during critical harvest times. Also, other or better markets may develop away from the original location.

Midwest Plan Service MWPS-2, *Farmstead Planning Handbook,* also discusses principles for locating grain storage and other buildings on a farmstead.

Electrical Service

Electrical service is a factor in site selection, but is usually not decisive. Grain centers put near the farmstead electrical service for convenience may soon outgrow the location and become unworkable. Space, traffic flow, and drainage are more important. Electrical service can be extended or a new service installed. However, sometimes a separate service means higher power costs. Discuss with your electrician and power supplier the costs for power extension and service to a better location. Base decisions on electrical and site preparation costs and expected future needs.

Get MWPS-28, *Farm Buildings Wiring Handbook,* for more information about materials and methods for electrical equipment and wiring in agricultural buildings.

Arrangement and Layout

In the plan layout, consider the arrangement of:

- Grain handling and cleaning equipment.
- Wet holding, drying, cooling, and storage bins.
- Dryer.
- Fuel storage.
- Scale for weighing.
- Management building.
- Feed processing and feed storage.
- Electrical service and control.
- Roads.

Arrange facilities to minimize walking, working outdoors, hand labor, and the need for brute strength.

A good rule is to keep it simple and compact with ample room for growth and mechanization.

Much of grain equipment operates automatically and continuously. Equipment failure, loss of electrical power, or overload shutdowns waste time. Minimize down time by arranging the system for easy visual and/or electric control monitoring.

Cleaning Equipment

Consider equipment arrangements for the following grain cleaning possibilities:

- Wet grain into the system.
- Grain from dryer to cooling bin or storage.
- Grain from cooling bins to storage.
- Grain from storage.

In a well designed system, one grain cleaner can perform several of these operations. Typically, about 2% fines by volume (about 1% by weight) is removed from grain by farm grain cleaners. If you plan to store screenings, include separate storages for wet, warm, and dry screenings. Wet or warm screenings have a very short storage life, are usually fed, and must be moved out often. Provide about 20 ft^3 of storage per 1,000 bu of grain processed with a grain cleaner.

Bins

Wet holding bins are usually filled directly from the receiving area and unload into a dryer. They are usually near the dryer and are often over it for gravity filling. See the section on wet grain holding in Chapter 3.

Grain storage. Allow for expanding storage in at least one and preferably two directions, Fig 5-2. Bins can be arranged in one or two rows to be filled with horizontal conveying equipment from the central grain handling facility. See Chapter 4.

Allow for overflow or emergency storage in your plan. High yields, changes in marketing strategy, or government storage and loan programs may require additional storage.

Plan for loading and unloading emergency storage. Consider a site near the grain center for easy filling. Delivery height to flat storages is usually low, so they can be farther from a bucket elevator. Pneumatic conveyors transport grain several hundred feet. Emergency storage is usually needed only 1 or 2 years out of 5, so hauling grain may allow better use of a structure the other years. Consider expansion when locating emergency storage.

Dryers

Dryers are typically near the system hub for easier conveying. Layout depends on the drying method, so consider present and future needs. As your operation expands, drying method or capacity may change. See Chapter 3 for discussion of different drying methods.

Changing the drying method is more difficult than expansion. For example, a bin dryer needs different space, arrangement, and grain handling than a self-contained dryer. Consider possible locations, positioning, and possibly altering the handling system with a change in drying method.

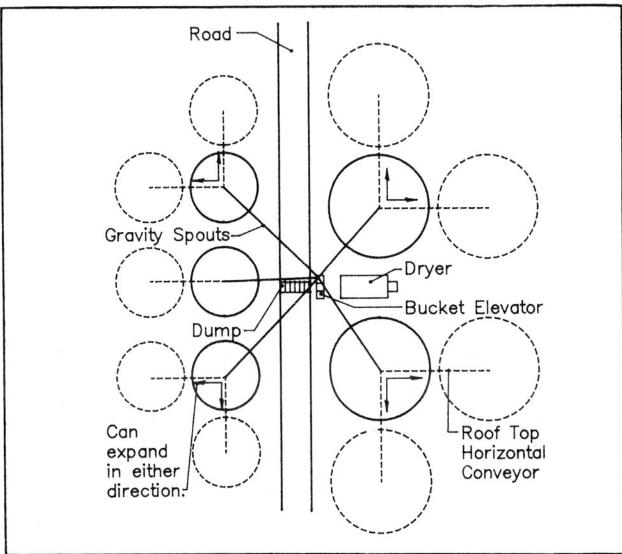

Fig 5-2. Storage expansion.
The dashed bins show possible expansion routes.

Fuel Storage

Large fuel storages reduce grain drying interruptions. A 250 to 300 bu/hr dryer may use 50 gal/hr of LP gas. Drying 30,000 bu of corn can require up to 6,000 gal of LP gas. Take advantage of price breaks often available during low use periods with a one-year storage.

Put fuel storage near grain receiving and drying. Provide easy access. Follow safety, fire code, and insurance requirements.

Alternate fuels, such as biomass, may become a major source of energy for grain drying. Plan space close to the dryer for storage and handling of these fuels. See Chapter 3.

Weigh Scale

Most farms do not have a scale certified for grain trading. However, include space to add a scale later for:

- Weighing grain in and out.
- Determining yields.
- Keeping loadout trucks under weight limits.
- Monitoring custom drying operations.
- Determining shares under a share-rent agreement.

Consider a scale in the grain receiving area, Fig 5-3. A scale in the receiving area eliminates moving vehicles between weighing and dumping and reduces walking from scale to operations center. The scale house or office can be used to keep weight, moisture content, and drying records together. A building over weighing and receiving protects the scale from weather, prolongs scale life, and eliminates the need for a separate scale house.

Plan the scale in line with and adjacent to the receiving dump. Separate the receiving dump edges and the scale pit by at least 3′ to prevent grain

Table 5-1. Electronic scale capacity.
A 60′ scale is needed to weigh semi-trailers, but is expensive and difficult to arrange in a layout. Shorter scales are often adequate for farm use.

Platform	Size	Total capacity lb	Capacity per section[a] lb
Wood	10′x20′	25,000	
	10′x30′	60,000	
	10′x40′	80,000	
Concrete	12′x24′	50,000	
	10′x30′	60,000	48,000[b]
	10′x40′	80,000	
	10′x60′	120,000	60,000
	10′x70′	120,000	60,000

[a]Capacity per section is for electronic scale platforms that are divided into two segments which can weigh separately or together for increased precision.
[b]The 10′x30′ unit can be divided into unequal segments for more utility. One segment handles 48,000 lb; the other 12,000 lb.

spillage into the scale pit. With a covered receiving area, put the scale pit at least 3′ from the door to minimize problems from blowing snow and rain.

A scale in the receiving lane may interfere with weighing grain for loadout when receiving volume exceeds 75,000 bu/yr or there are unusual weighing needs. Locate the scale away from the receiving area if marketing activity, livestock feed weighing, or custom grain drying conflict with receiving.

A scale remote from receiving works best when the volume handled justifies a full-time person at the scale house. On large farms, the scale house can be part of a farm office.

With separate receiving and weighing areas, put the scale:

- At least one vehicle length from the dump to weigh one vehicle while dumping another, Fig 5-3b.
- Along the vehicle entry-exit route for easy access. Allow traffic to flow both ways across the scale.
- With the scale beam or dial on the drivers′ side, so they can weigh their own loads.

If there is no permanent scale immediately planned, consider a portable, one-axle scale to check truck wheel weights for road restrictions.

Management Building

A building near the grain center can house controls, grain testing equipment, and records. Provide for moisture testing equipment, grading screens, desk, file cabinets, communications equipment, computer, coffeepot, and a cot. This building can also store portable equipment like sweep augers. The management building should be insulated and heated. The management building can be a portable building, trailer, or room in a larger building that includes grain receiving, feed processing, and load-out.

Feed Processing

Even if you do not currently need feed processing, plan for it. A cash grain farm today can have livestock tomorrow. Planning for feed processing increases flexibility and value to the next generation and prospective buyers. Feed processing equipment selec-

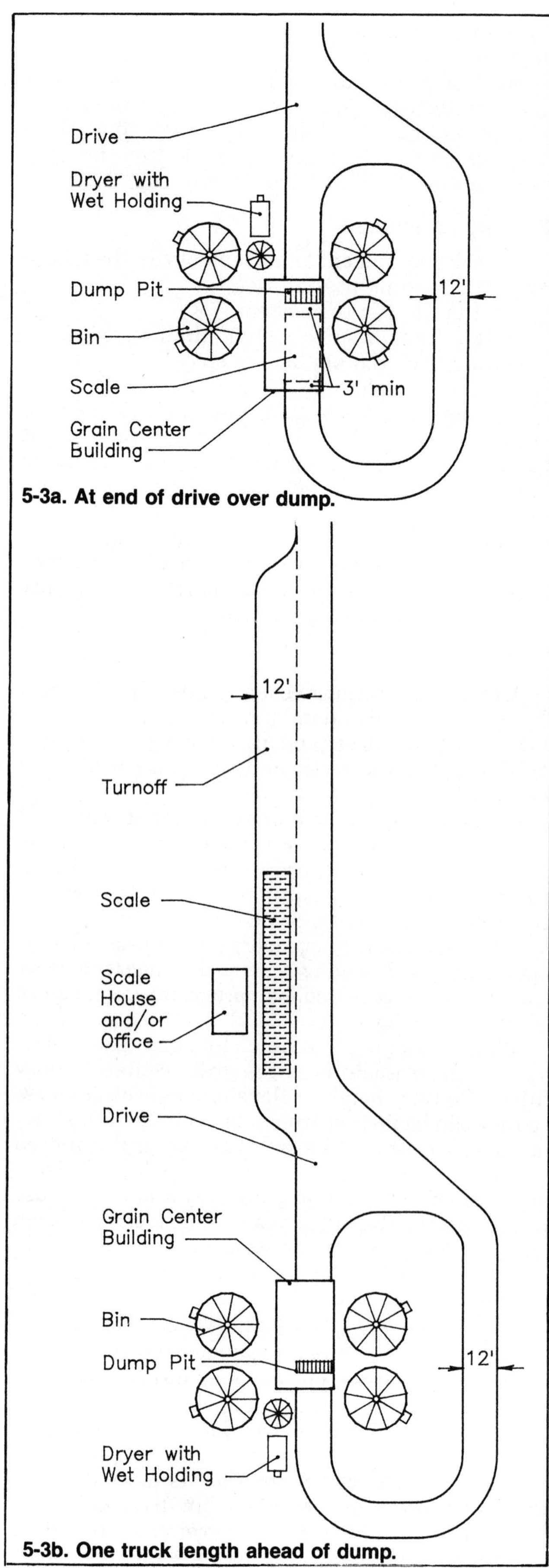

Fig 5-3. Scale locations.

tion, arrangement, and location is complex and is not included in this book. However, consider feed processing and delivery while planning.

Feed processing on small to medium livestock farms often starts with a portable grinder-mixer, which is competitive for feed volumes up to 400 tons/yr. Portable grinder-mixers can be moved between sites to collect grain and feed ingredients, but it is best to bring the grain to the mill.

Most grain centers with an operation center building can include space for a grinder-mixer and tractor in the grain receiving area. Overhead gravity flow bins, a weigh hopper or buggy to pre-weigh feed ingredients, and a storage area for base mixes and feed additives can also be incorporated. Use the same receiving and loadout facilities. Develop a feed processing area or building near the receiving area for convenient handling of bulk grain and feed and sacked or boxed microingredients. Provide a floor area of 8'x8' or more. This space can be used for storage until feed processing is needed.

Feed processing in the grain center works well on farms producing up to 100,000 bu of grain per year and feeding at least 400 tons per year.

If feed processing volume increases or the portable grinder-mixer wears out, consider a stationary grinder/mixer. With an operation center, it fits in the same space as the portable unit for an easy transition. If the grain or livestock operations increase to highly intensified, large volume enterprises, consider separating the grain center and feed processing. Stationary feed processing can be near the grain center or at a satellite location.

Satellite facilities are best for large operations to reduce traffic and space conflicts. With satellite facilities, livestock and grain operations are separate, except for replenishing the feed center. Plan a three week to three month supply of feed ingredients at the feed processing center. Up to a year's supply of feed ingredients can be stored at the feed processing satellite. However, small storages provide for more rapid draw-down, easier grain use estimates, and easier grain quality monitoring. With smaller bins, less special filling equipment is needed and aeration is less critical.

For new bins at a satellite facility, consider 6,000 to 10,000 bu capacity to keep cost per bu low. An existing 14'-24' bin that is too small for the grain center may be ideal for the feed processing center. An inclined portable conveyor can be left in position all year or a short vertical bucket elevator can be installed to fill the bins.

Electrical Service

Locate the grain center service entrance and controls panel near the receiving, drying, and loadout facilities. Avoid the high humidity discharge air from the dryer. Put the panel and controls in a building because weatherproof switches are more expensive than non-weatherproof switches. Consider locating them in the operations center building.

Put the controls where the operator can monitor them. Visual monitoring of grain flow at receiving

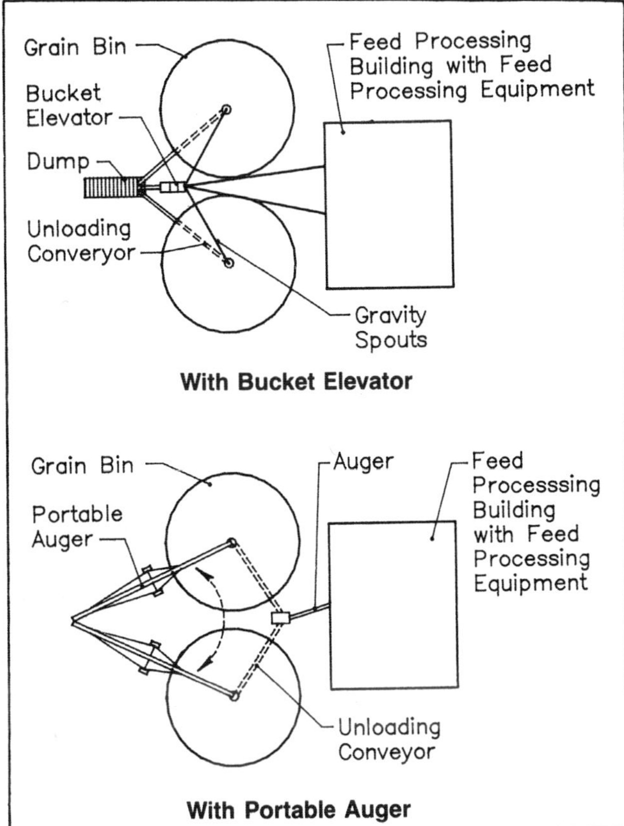

Fig 5-4. Example feed processing satellite facility.

and loadout are the most important. Install pressure and level sensors, warning lights, and/or horns to alert the operator of grain flow problems.

Grain center electrical requirements are large and can be complex. Plan for all equipment in the final system even though some tractor powered equipment is used initially. Consult your local power supplier during planning to determine electrical power needs and limitations. Discuss the possibility of immediate or future conversion to three-phase power. If future conversion to three phase is not feasible, consider phase converters or soft start motors for loads larger than 7½ hp. If three-phase service will soon be available, use phase converters and three-phase motors so the system can be easily converted. For convenience and safety, use underground wiring to and within the grain center.

Roads

Provide all-weather roads. Plan adequate space for snow removal equipment and snow storage.

Provide smooth traffic flow through and around all buildings and lots. Allow for expansion when locating and arranging roads. Farmstead traffic may not seem important, but as grain volume increases, transport vehicles get larger, and loading time increases, traffic delays can become a problem.

Plan straight access roads from the main road to the receiving/loadout area for easier maneuvering of loaded trucks. If needed, access roads can curve, but plan all turns and loops properly. See Fig 5-5 for road curve dimensions.

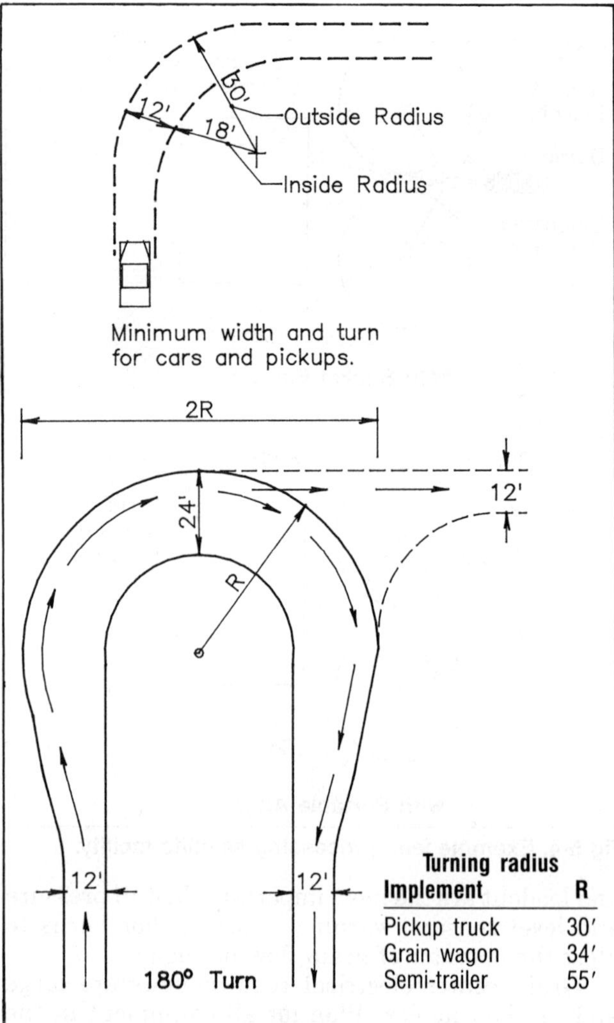

Turning radius	
Implement	R
Pickup truck	30′
Grain wagon	34′
Semi-trailer	55′

Fig 5-5. Road curve dimensions.

Straight access roads cannot be screened from public view. If desired, offset the road one width and plant shrubs or trees to block the view. If the grain center is close to the road or in the corner of the intersection of two roads, consider more than one access road.

In your layout, plan roads so all vehicles drive through the receiving and loadout area. Backing up to unload or load can be tolerated, but it is inconvenient, unsafe, and nearly impossible for high volume grain handling.

Developing Your Plan

Rough Sketching

Rough sketching is the best way to start developing a plan. Only a pencil, paper, eraser, and straight edge are needed. Sheets of 8½″x11″ plain or graph paper are easy to work with and are large enough to draw preliminary sketches for evaluation.

The basic idea of rough sketching is to quickly draw a few alternative plans. Include future needs such as feed processing, a scale, a grain cleaner, and emergency storage. The drawings should resemble actual equipment, but need not be detailed.

Do not spend a lot of time drawing facilities precisely. Measure only enough to be close. You will learn more from comparing 4 or 5 rapidly sketched alternatives than from 1 or 2 neat drawings that limit your options. Neat, precise drawing is done **last.**

Rough sketching is a trial and error process. There are no rules or equations to easily develop an arrangement. It takes time and imagination. Rough sketching is a productive and relatively easy way to start, but can also be frustrating. Many ideas may be sketched before one fits the guidelines of this book and specific farmstead grain handling and family needs.

Start by drawing an arrow indicating direction. North is usually at the top of the sheet. Put in any existing facilities. Then, locate the future drive and show its direction. Arrange facilities around the drive. Many grain centers start with a dryer and a few storage bins. Storages are usually in 1 or 2 rows along the drive. A bucket elevator is usually located beside the drive and the dump pit is usually in the drive.

Another approach is to use cut outs of major items. Arrange them on a piece of paper with the future drive drawn on it. When you select a layout you like, draw around the cutouts or place tracing paper over them and draw them.

Try to use existing or near-future facilities. For example, a drying bin now can be a cooling or dryeration bin later and auger-filled storages now can receive grain from a bucket elevator later. Self-contained dryers and wet holding bins can be relocated as the grain center develops.

After completing the rough sketches, make a photo copy and add arrows to show grain flow from grain receiving to the dryer, from the dryer to storage, etc. Work toward grain flowing from the receiving/loadout hub to any component and back again.

Sketches of grain drying, handling, and storage systems are in Figs 5-7 to 5-12. The symbols used in Fig 5-7 to 5-12 are defined in Fig 5-6. **These are examples only,** to show how different grain drying, conditioning, and handling methods can be organized into well planned systems. They are patterns for planning—not all of these ideas will fit your farm or needs, but the basic ideas and principles can be applied.

Scale Drawing

After deciding on the best one or two layouts, draw a top view scale drawing of the equipment and buildings. Figs 5-7 to 5-12 show scale drawings of existing and near-future equipment for each rough sketch. Refer to Fig 5-6 for symbol definition.

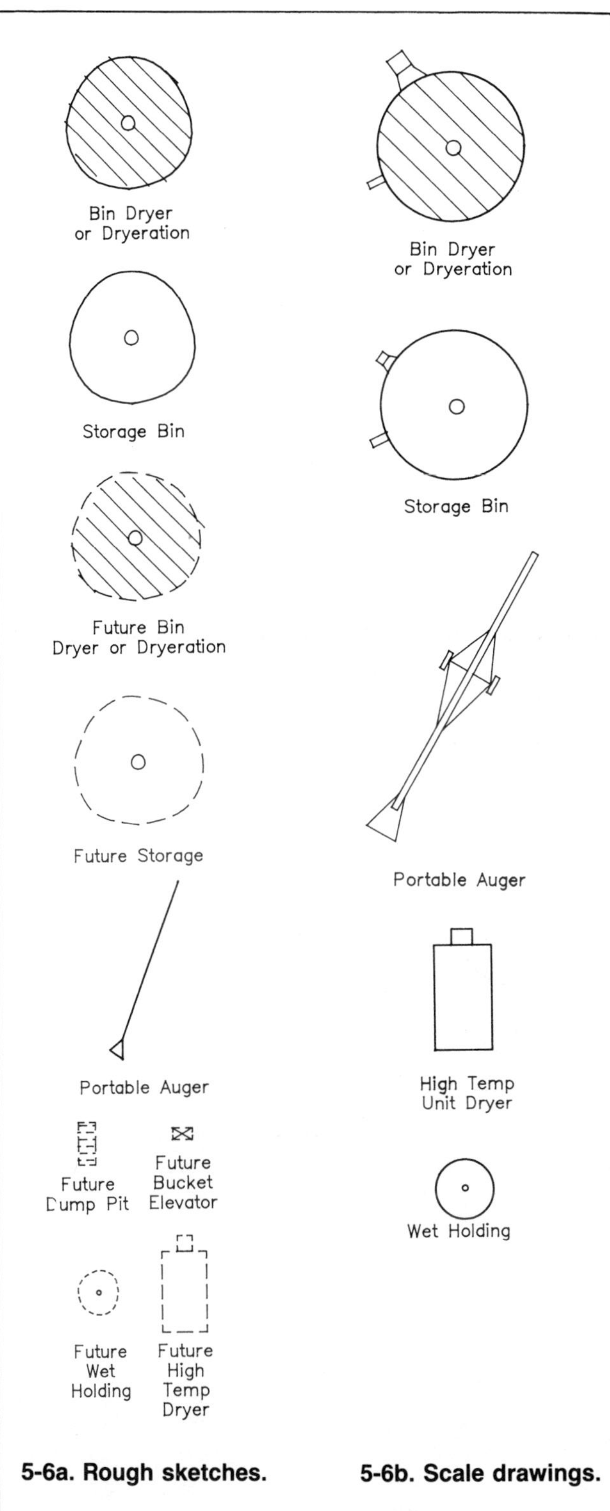

5-6a. Rough sketches.

5-6b. Scale drawings.

Fig 5-6. Legend for Figs 5-7 to 5-12.

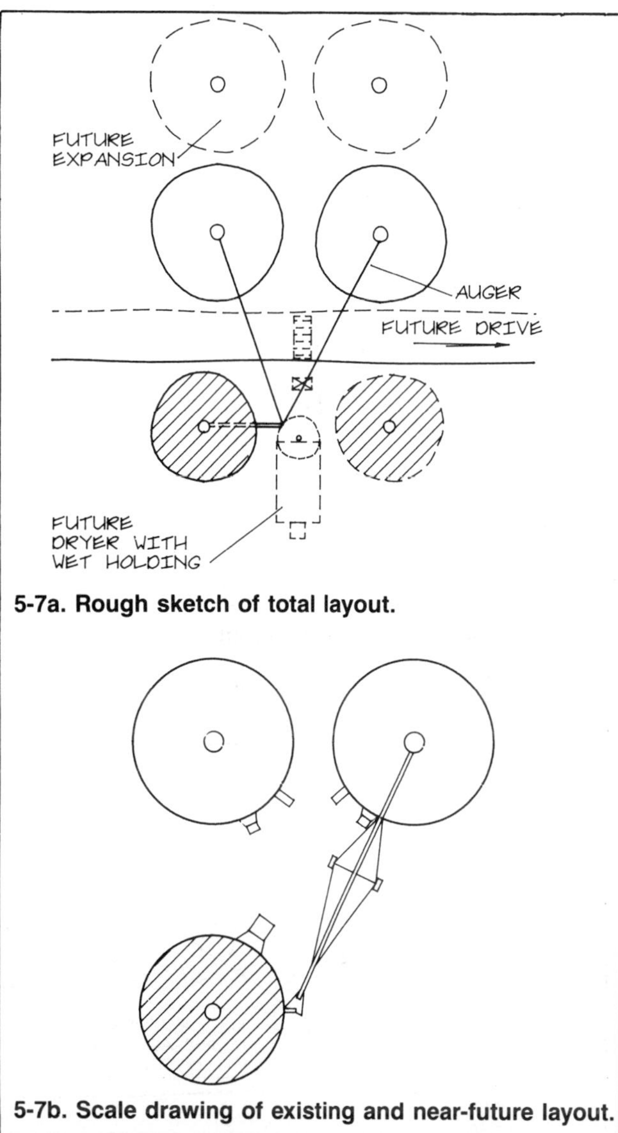

5-7a. Rough sketch of total layout.

5-7b. Scale drawing of existing and near-future layout.

Fig 5-7. Bin dryer, bottom unloading, future dryer perpendicular.
Future expansion from an existing drying bin.

Graph paper is recommended to make scale drawings. Use 11″x17″ sheets to include more detail. Use a circle template or compass to draw circular bins and a ruler or scale to draw the plan accurately. Pick a scale and record it on your drawing. For example, on graph paper, one square might equal 2′. If a ruler is used, different scales are available—1/16th, 1/8th, or 1/10th.

Equipment on the rough sketch might not fit when drawn to scale. It may be necessary to relocate equipment several times before things fit together properly.

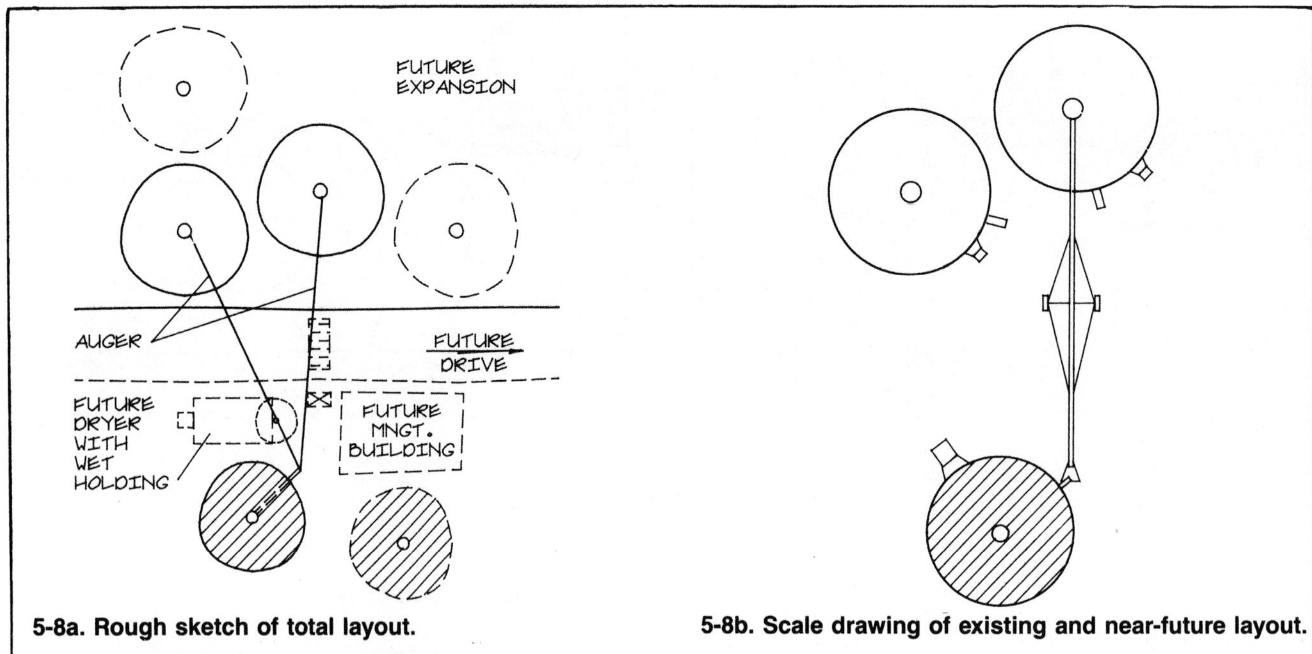

5-8a. Rough sketch of total layout.

5-8b. Scale drawing of existing and near-future layout.

Fig 5-8. Bin dryer, bottom unloading with dryer and management building.

TOP UNLOADING AUGER

AUGER

FUTURE DRYER
WITH WET HOLDING

FUTURE
MANAGEMENT
BUILDING

FUTURE DRIVE

FUTURE EXPANSION

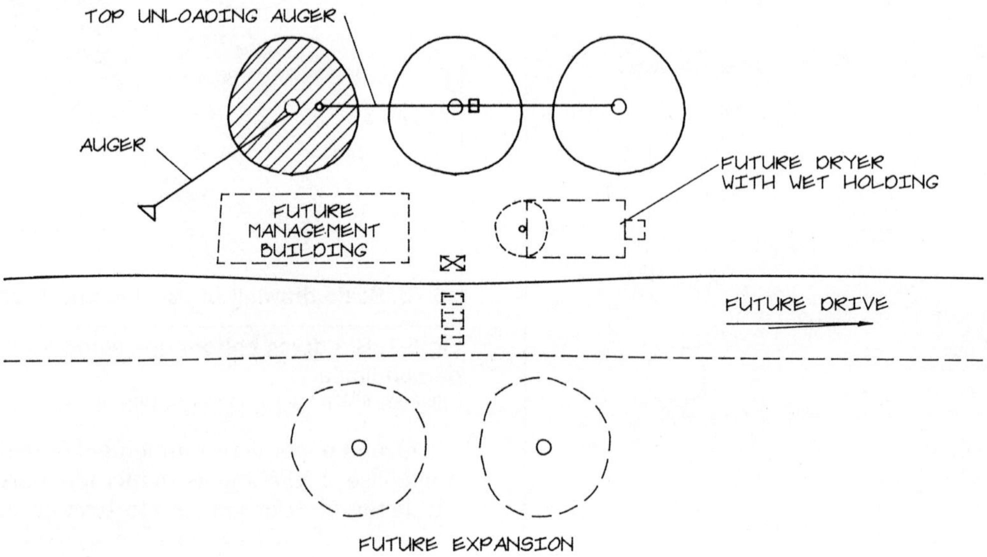

5-9a. Rough sketch of total layout.

Top Unloading Auger

Auger

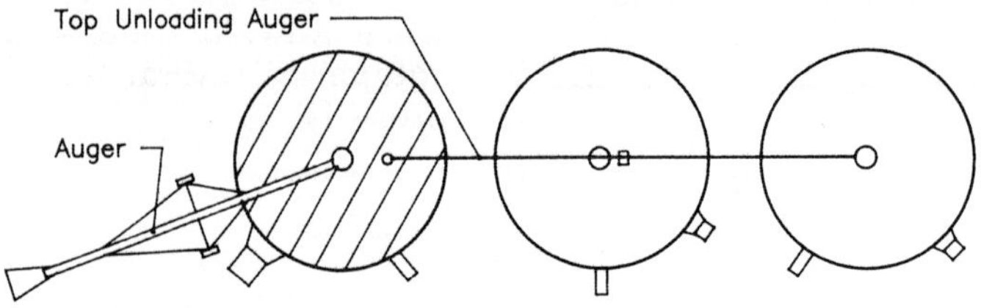

5-9b. Scale drawing of existing and near-future layout.

Fig 5-9. Continuous flow bin dryer, future dryer and management building.

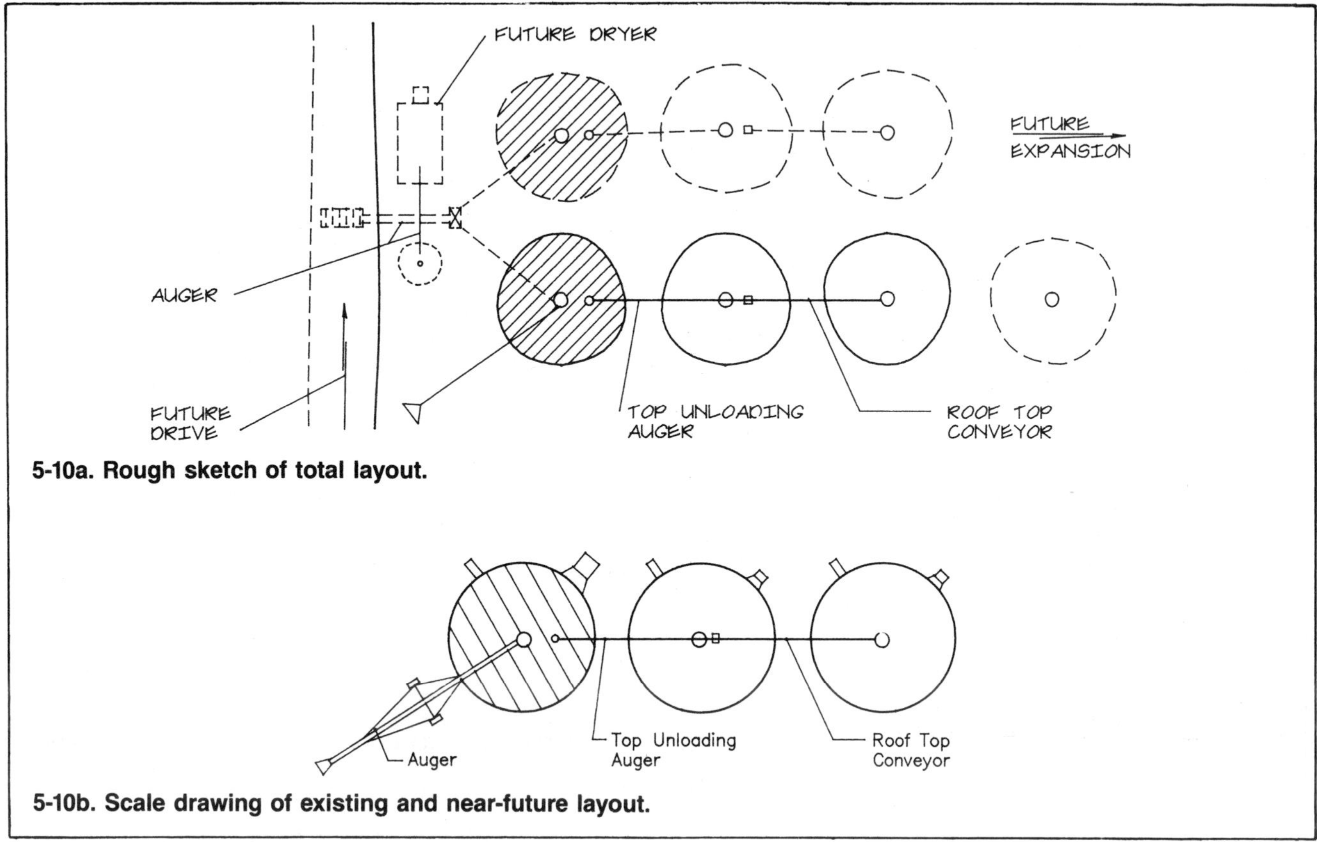

5-10a. Rough sketch of total layout.

5-10b. Scale drawing of existing and near-future layout.

Fig 5-10. Continuous flow bin dryer, top unloading.

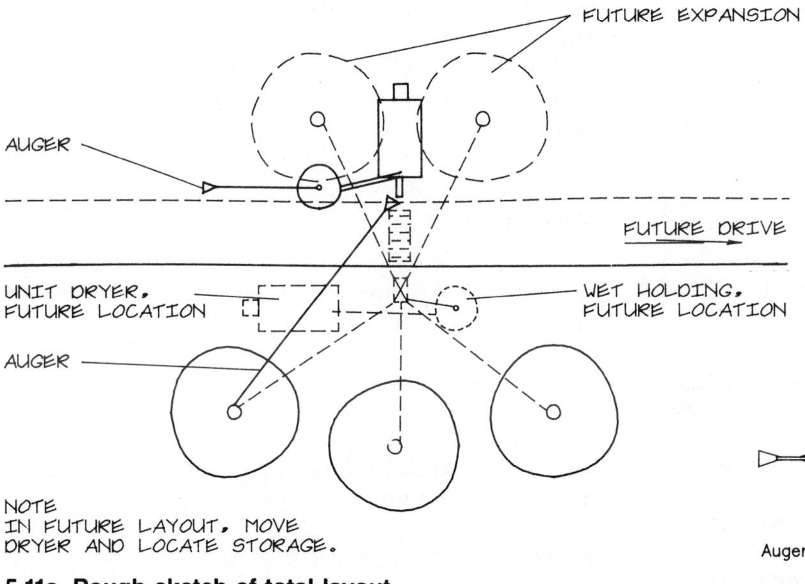

5-11a. Rough sketch of total layout.

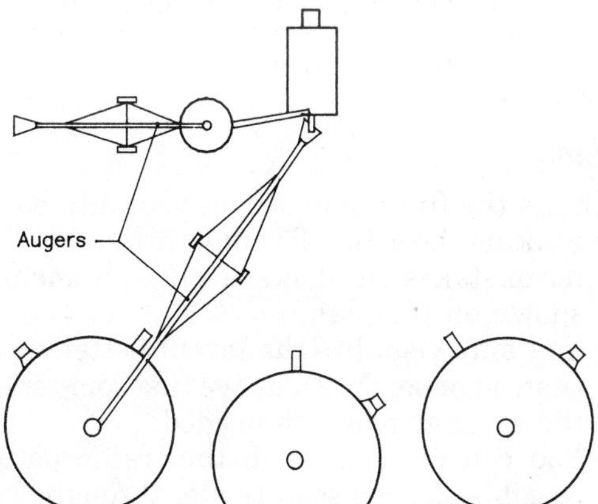

5-11b. Scale drawing of existing and near-future layout.

Fig 5-11. Self-contained unit dryer, curved layout.
Continuous flow or batch dryers. Existing self-contained, high temperature dryer and auger can be moved when expanding to a central operations center and bucket elevator.

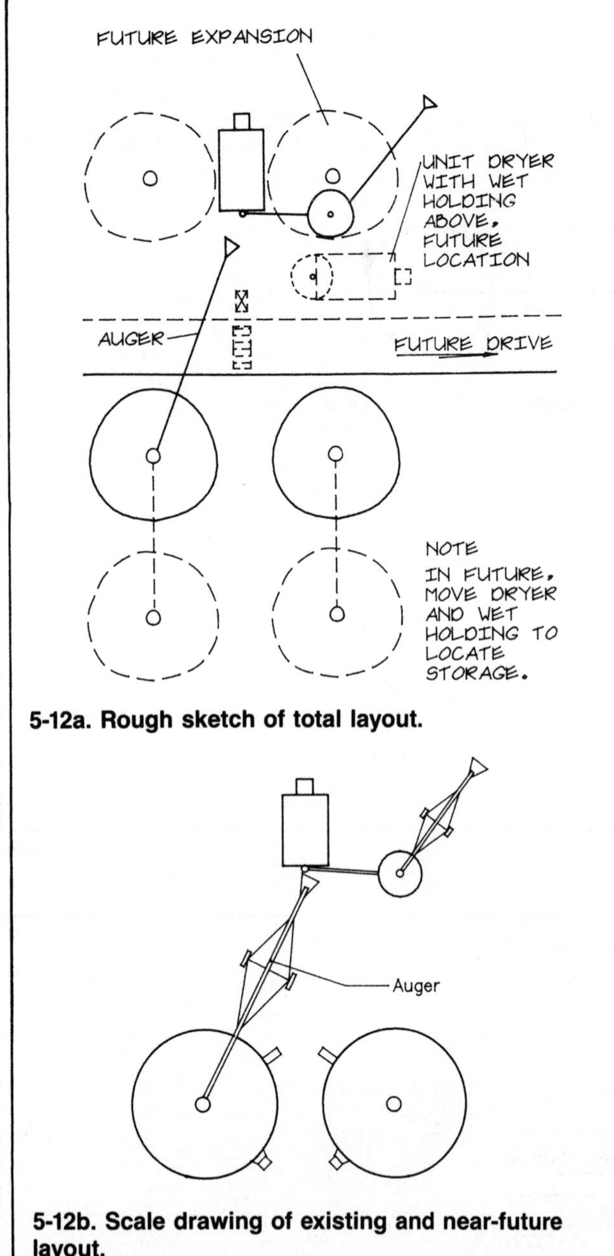

5-12a. Rough sketch of total layout.

5-12b. Scale drawing of existing and near-future layout.

Fig 5-12. Self-contained unit dryer, linear bin layout.
Continuous flow or batch dryers. Existing self-contained, high temperature dryer and auger can be moved when expanding to a central operations center and bucket elevator.

Staking

Stake the final plan out on your site because:

- Staking the actual layout helps ensure there are no mistakes. It shows if all equipment fits as shown on the plan.
- You may visualize the layout better on the site than on paper. You may see that some changes in the arrangement are needed.
- You can drive through the traffic patterns. If possible, drive a semi-trailer through because it is difficult to maneuver in tight or improperly arranged traffic ways.
- It allows you to evaluate slope and drainage.

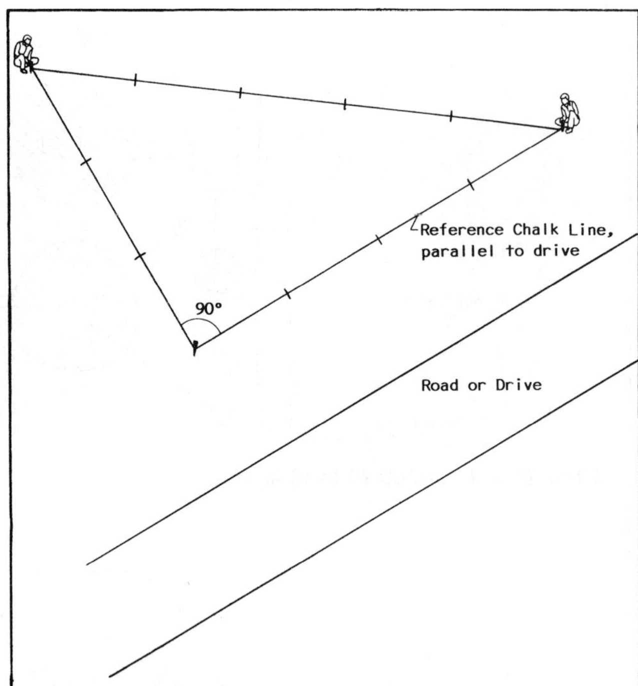

Fig 5-13. Staking a 90° angle with a 3-4-5 triangle.
A 90° angle is formed between the two shorter sides in any triangle where the side proportions are 3, 4, and 5. Example combinations are: 3′-4′-5′; 9′-12′-15′; or 30′-40′-50′.

Two people can stake a layout with a 100′ measuring tape, chalkline, and some stakes or flags. First, locate the future drive or other limiting boundary (road, tree line, or existing facility) as a reference for locating equipment. Use 3-4-5 triangles to form 90° angles, Fig 5-13.

Mark the outline of each piece of equipment with lime or fertilizer. Do not forget to include future developments, such as expansion or modification of grain drying and storage; feed processing; grain cleaning; and scales.

Evaluate the Plan

Evaluate your plan by going through the checklist in Table 5-2. All answers need not be "excellent" for an acceptable system, but do not accept any "poor" answers. Consider each question carefully. Do not make an error in judgment today that you may regret in the future. Use the checklist to identify both small and potentially serious problems. It is less costly and easier to correct a mistake during planning than after construction.

Selecting Equipment Suppliers and Contractors

Initial cost is important, but selecting reliable builders and suppliers that promptly service equipment when required is even more important.

Often purchases are based on initial price only. However, a more expensive drying bin, installed auger system, or other equipment that performs well is better than less expensive equipment that performs poorly or breaks down frequently.

Table 5-2. Grain systems checklist.

Excellent	Fair	Poor	
☐	☐	☐	The grain center site includes space for expansion and future improvements.
☐	☐	☐	There are no surface or subsurface water problems at the site.
☐	☐	☐	All-weather roads are planned for year-round access.
☐	☐	☐	Safety practices and considerations are recognized and will be implemented.
☐	☐	☐	A management building is planned to house electrical controls, grain testing equipment, etc.
☐	☐	☐	Additional storage will have good grain handling connections.
☐	☐	☐	Electric power requirements have been discussed with the power supplier.
☐	☐	☐	Grain center is convenient to farmstead for monitoring, safety, and theft protection.
☐	☐	☐	All equipment can be easily maintained.
☐	☐	☐	Fan noise and dust will not be a problem to the farm home or neighbors.
☐	☐	☐	Wet grain receiving is fast and convenient. Grain moves conveniently between locations in the system.
☐	☐	☐	All bins and buildings are planned to be built on concrete floors at least 12″-16″ above grade.
☐	☐	☐	The grain center is at least 200′ from the farm home.
☐	☐	☐	Adequate storage space and convenient access are planned for LP gas, other fuels, and possibly biomass.
☐	☐	☐	Roads and drives can handle or be adapted for semi-trailer trucks and commercial grain traffic.
☐	☐	☐	Improved drying methods such as dryeration, in-bin cooling, combination or natural air drying can be easily adopted.
☐	☐	☐	Grain cleaning equipment is included or can be added.
☐	☐	☐	A farm scale can be easily included in the grain center.
☐	☐	☐	Space is planned to build a feed processing and delivery system.
☐	☐	☐	The grain center is conveniently located in relation to present or future livestock facilities.
☐	☐	☐	The grain center is located close to the middle of the operation which is about the same distance from all cropland.
☐	☐	☐	Equipment is oriented to take advantage of the winter sun.

Find out how long equipment suppliers and contractors have been in business. Generally, those in business for many years have done good work and promptly provided service. Local dealers and contractors often want to do a good job because they want future business in the area.

Check the performance of an equipment supplier or contractor with other farmers. Get a list of several people with whom the salesperson, company, or contractor has dealt. Select 2 or 3 operations that are similar to yours and contact them for their opinion.

Problems often result from poor communication. Verbal communication is easy to misinterpret. Put the agreement in writing and have it signed and dated. Too often the only signed and dated paper is the invoice of equipment to be supplied and the price. Some other important points to be agreed on in writing are:

- Equipment will be installed according to manufacturer's specifications. Equipment is guaranteed to operate as specified by the salesperson.

Examples:

a. A dryer will dry a specified number of bushels of grain per hour removing a specified amount of moisture under specified ambient air conditions.
b. Conveyors will move a specified number of bushels of grain per hour.
c. Each conveyor in a conveying series will move 5%-10% more grain than the previous conveyor to minimize plugging and grain damage.
d. Drying, cooling, and aeration fans will deliver a specified airflow rate (cfm/bu) at a given static pressure.

Before construction starts, also agree in writing:

- All construction will be completed and equipment installed, in working condition, by a specified date.
- A payment procedure is specified in the contract.
- Disputes will be resolved by a mutually agreed upon arbitrator.

- Remove at least 6″ of topsoil before placing any earth or sand fill to reduce uneven settling. Pack earth fill with a sheeps-foot roller or equivalent.
- Place concrete on a packed, wetted surface. Put a plastic vapor barrier under concrete.
- Make concrete bin floors, unless otherwise agreed upon, at least 12″ above grade. Make building floors and concrete pads under dryers, fans, and hopper bottom bins at least 6″ above grade.
- Reinforce concrete with welded wire to reduce cracks from uneven settling and to prevent cracks from opening wider.
- Agree upon the mix, curing procedures, and finishes (troweled, broomed, floated, etc.) of all concrete floors.

Plans, Specifications, and Contracts

Detailed documents help provide needed communication and understanding between owner, builder, and lender. **Plans** show all necessary dimensions and details for construction. **Specifications** support the plans; they describe the materials to be used, including size and quality, and often outline procedures for construction and quality of workmanship. The **contract** is an agreement between the builder and the owner; it includes price of construction, schedule of payments, guarantees, responsibilities, and starting and completion dates.

You have several options for preparing plans and specifications.

Some dimensioned plans are available from the Extension Agricultural Engineer at any of the institutions on the inside front cover or the Midwest Plan Service. See Chapter 8 for available plans and publications.

Be your own contractor. Draw a final plan including the specifications. Make sure dimensions are correct and construction details and materials are determined. Have plans checked by the appropriate regulatory agency when required. Determine total costs before beginning construction.

Use a design and construction firm. Some firms make working drawings and specifications and have standard contract forms.

Hire a consulting engineer.

Consulting Engineers

Services commonly offered by consulting engineers include:

- Direct personal service (technical advice, etc.).
- Preliminary investigations, feasibility studies, and economic comparison of alternatives.
- Design.
- Cost estimates.
- Engineering appraisals.
- Bid letting.
- Construction supervision and inspection.

Consulting engineers usually do a project in three phases: preliminary planning, engineering design, and construction monitoring. They may be retained to help with one or more of these phases. To select a consulting engineer, consider:

- Registration: to protect the public welfare, states certify and license engineers of proven competence. Practicing consulting engineers must be registered professional engineers in their state of residence, and qualified to obtain registration in other states where their services are required.
- Technical qualifications.
- Reputation with previous clients.
- Experience on similar projects.
- Availability for the project.

6. APPENDIX

Conveyor Distances

Table 6-1. Inclined conveyor discharge heights and distances.
Values are approximate. See Fig 6-1 for variable definitions.

	Conveyor angle					
	25°		35°		45°	
L	H	D	H	D	H	D
ft						
21	9	19	12	17	15	15
26	11	23½	15	21	18½	18½
31	13	28	18	25	22	22
36	15	32½	21	29	25½	25½
41	17½	37	23½	33½	29	29
46	19½	41½	26½	37½	32½	32½
51	21½	46	29½	41½	36	36
56	23½	51	32	46	39½	39½
61	26	55	35	50	43	43
66	28	60	38	54	46½	46½
71	30	64½	40½	58½	50	50

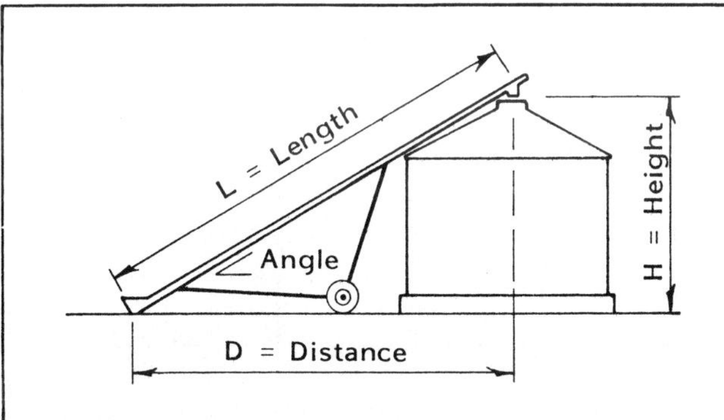

Fig 6-1. Inclined conveyor definition.

Table 6-2. Minimum gravity spout angles and floor slopes.
Values are absolute minimums for grain flow. Varnishing, waxing, or polishing wood and steel floors reduces friction and improves flow. If spouting is rusty, bulky or damp, grain will flow better if spouting is scoured with dry hard-finish grain. If flow stops, it is difficult to restart.

Material	Spout angle or floor slope	Slope equivalents rise:run
Grains, dry	37°	¾:1
Grains, wet, hot, or tough	45°	1:1
Meal	60°	1¾:1
Pellets	45°	1:1
Sunflower	60°	1¾:1

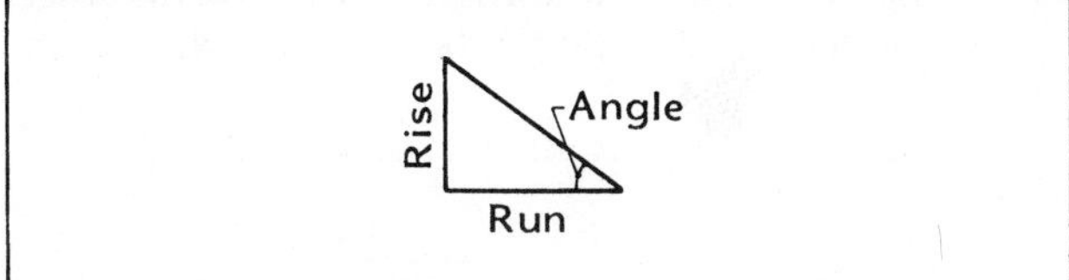

Fig 6-2. Slope triangle.

Table 6-3. Gravity spout heights and lengths.
See Fig 6-3 for variable definitions.

Distance to bin center	Spout angle					
	37°		45°		60°	
D	L	H	L	H	L	H
ft						
16	20	12	23	16	32	28
18	22½	13½	25½	18	36	31
20	25	18	28½	20	40	35
22	27½	16½	31	22	44	38
24	30	18	34	24	48	42
26	32½	19½	37	26	52	45
28	35	21	40	28	56	48½
30	37½	22½	42½	30	60	52
35	44	26½	49½	35	70	61
40	50	30	57	40	80	70
45	57	34	64	45	90	78
50	63	38	71	50	100	87

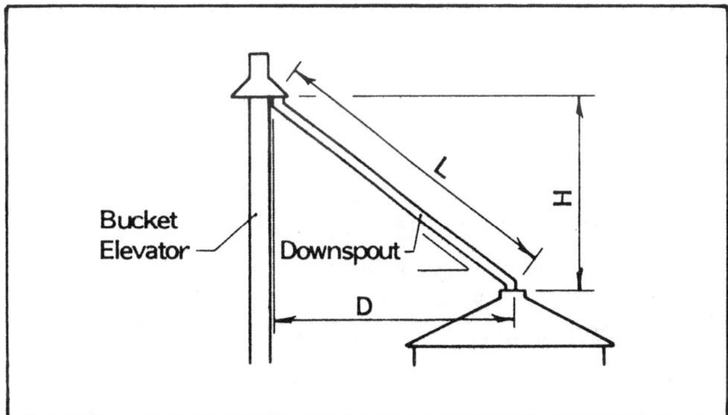

Fig 6-3. Gravity spout definition.

Inclined conveyor example

Example 6-1:

Determine the auger length and horizontal distance between an auger hopper and bin center if the vertical discharge height is 25′and the auger is inclined 35°. See Fig 6-1.

Solution:

Use Fig 6-4. For inclined conveyors: the vertical distance is the discharge height, horizontal distance is the distance between the auger hopper and bin hatch, and slope distance is the auger length. From Fig 6-4, a 44′ long auger is needed and the hopper end is 36′ from the center of the bin.

Gravity spout examples

Example 6-2:

Determine the elevator discharge height and spout length for a gravity spout at a 60° incline discharging to a bin 20′ from the bucket elevator.

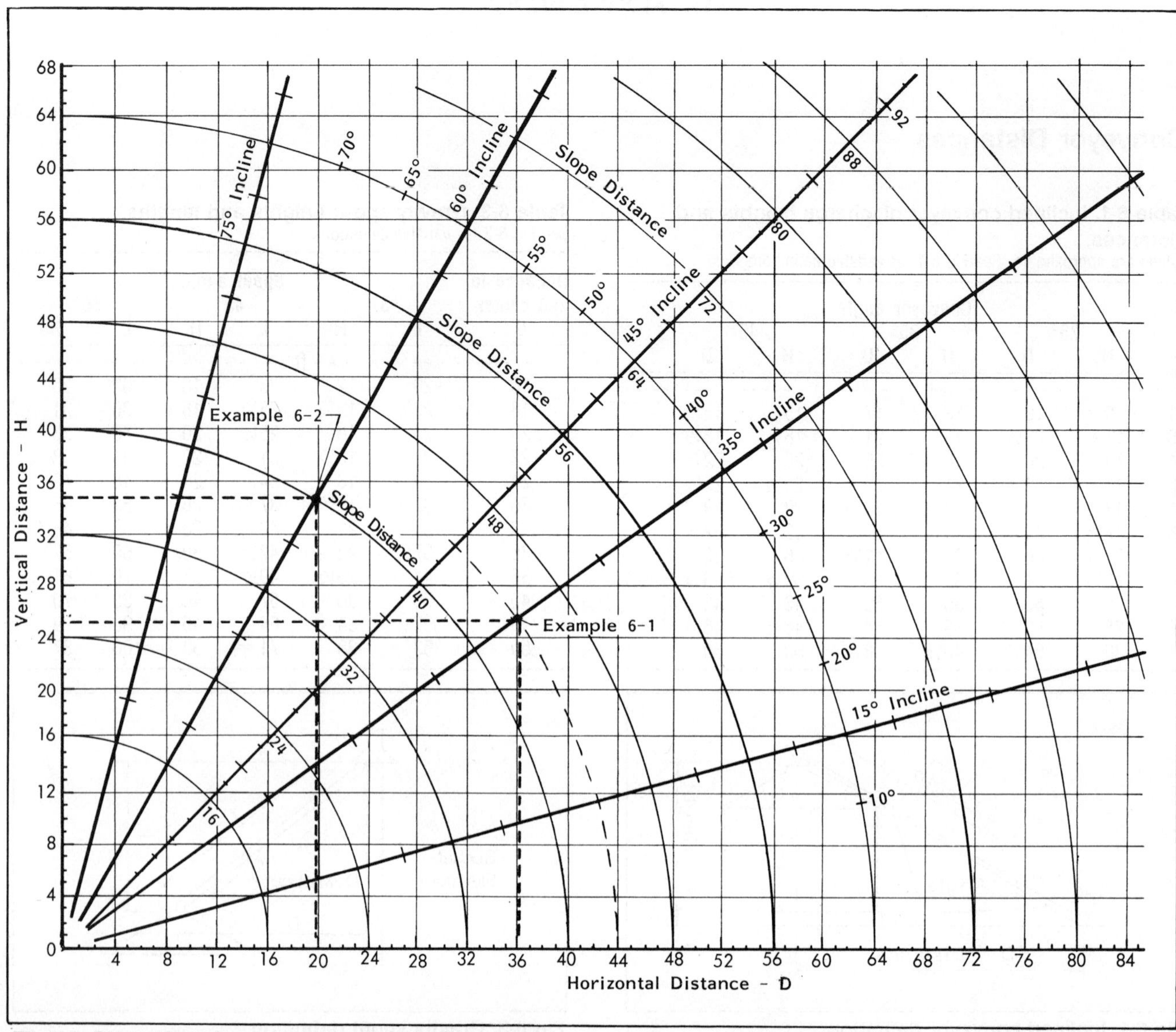

Fig 6-4. Slope chart for conveyors and gravity spouts.
Chart variables are vertical distance or rise, V; horizontal distance or run, H; conveyor length, L; and conveyor incline, I. Use any two variables to determine the other two. For example, if conveyor length and slope are known, then vertical and horizontal distance can be determined.

Solution:

a. Use Fig 6-4. For gravity spouting: vertical distance is the discharge height above the bin top, horizontal distance is the distance between the bucket elevator and bin top, and slope distance is the spout length. See Fig 6-3.

   From Fig 6-4, the elevator discharge must be 35′ above the bin top and the spout is 40′ long.

b. Using Table 6-3: spout angle = 60° and distance to bin center, D = 20′.

   The discharge height above the bin top is 35′ and spout length is 40′.

Example 6-3:

Determine the vertical discharge height for a gravity spout at a 37° incline discharging dry grain from a bucket elevator to a bin top 24′ away.

Solution:

a. Using Table 6-2, the slope equivalent (rise:run) is ¾:1. For this case, run = 24′; determine rise.
   Rise = ¾ × Run = ¾ × 24′ = 18′
   The vertical discharge height between the bucket elevator discharge and bin top is 18′.

b. Using Table 6-3: spout angle = 37° and distance to bin center, D = 24′, determine the discharge height, H.
   From Table 6-3, the discharge height is 18′.

Storage Volume Tables

Table 6-4. Grain seed densities.

Based on weights and measures used by the U.S. Department of Agriculture. Adapted from ASAE Data D241.2, 1984.

Grain or seed	Unit	Approximate net weight[a] lb	Bulk density lb/ft³	Grain or seed	Unit	Approximate net weight,[a] lb	Bulk density lb/ft³
Alfalfa	bushel	60	48.0	Oats	bushel	32	25.6
Barley	"	48	38.4	Orchard grass	"	14	11.2
Beans:				Peanuts, unshelled:			
Lima, dry	"	56	44.8	Virginia type	"	17	13.6
Lima, unshelled	"	32	25.6	Runners, Southeastern	"	21	16.8
Snap	"	30	24.0	Spanish			
Other, dry	"	60	48.0	Southeastern	"	25	19.7
(navy, pinto, etc.)				Southwestern	"	25	19.8
Bluegrass	bushel	14-30	11.2-24.0	Perilla	"	37-40	29.6-32.0
Broomcorn	"	44-50	35.2-40.0	Popcorn:			
Buckwheat	"	48-52	38.4-41.6	On ear	"	70[b]	28.0
Castor beans	"	46	36.8	Shelled	"	56	44.8
Clover	"	60	48.0	Poppy	"	46	36.8
Corn:				Rapeseed	"	50 & 60	40.0 & 48.0
Ear, husked	"	70[b]	28.0	Redtop	"	50 & 60	40.0 & 48.0
Shelled	"	56	44.8	Rice, rough	bushel	45	36.0
Green sweet	"	35	28.0	Rye	bushel	56	44.8
Cottonseed	"	32	25.6	Sesame	"	46	36.8
Cowpeas	"	60	48.0	Sorgo	"	50	40.0
Flaxseed	"	56	44.8	Soybeans	"	60	48.0
Grain sorghums	"	56 & 50	44.8 & 40.0	Spelt (p. wheat)	"	40	32.0
Hempseed	"	44	35.2	Sudan grass	"	40	32.0
Hungarian millet	"	48 & 50	38.4 & 40.0	Sunflower, non-oil	"	24	19.2
Kafir	"	56 & 50	44.8 & 40.0	Sunflower, oil	"	32	25.6
Kapok	"	35-40	28.0-32.0	Timothy	"	45	36.0
Lentils	"	60	48.0	Velvet beans (hulled)	"	60	48.0
Millet	"	48-50	38.4-40.0	Vetch	"	60	48.0
Mustard	"	58-60	46.4-48.0	Wheat	"	60	48.0

[a]Net weight is the standard bushel weight if one has been defined.

[b]70 lb of husked, ear corn yields 1 bu, or 56 lb of shelled corn. 70 lb ear corn occupies 2 volume bushels (2.5 ft³).

Table 6-5. Capacities of level full round bins.

Based on 1.245 ft³ = 1 bu. Weight of bin contents varies by actual material density, packing, and bin surface configuration.

Grain depth ft	Bin dia., ft 15	18	21	24	27	30	33	36	40	42	44	48	60
	bushels												
1	142	204	278	363	460	568	687	818	1,009	1,113	1,221	1,453	2,271
10	1,419	2,044	2,782	3,634	4,599	5,678	6,870	8,176	10,093	11,128	12,213	14,535	22,710
12	1,703	2,453	3,338	4,360	5,519	6,813	8,244	9,811	12,112	13,354	14,656	17,441	27,252
14	1,987	2,861	3,895	5,087	6,438	7,949	9,618	11,446	14,131	15,579	17,098	20,348	31,794
16	2,271	3,270	4,451	5,814	7,358	9,084	10,992	13,081	16,150	17,805	19,541	23,255	36,336
18	2,556	3,679	5,008	6,541	8,278	10,220	12,366	14,716	18,168	20,030	21,984	26,162	40,878
20	2,839	4,088	5,564	7,267	9,198	11,355	13,740	16,351	20,187	22,256	24,426	29,069	45,420
22	3,123	4,497	6,120	7,994	10,117	12,491	15,114	17,987	22,206	24,482	26,869	31,976	49,963
24	3,407	4,905	6,677	8,721	11,037	13,626	16,488	19,622	24,224	26,707	29,311	34,883	54,505
26	3,690	5,314	7,233	9,447	11,957	14,762	17,862	21,257	26,243	28.933	31,754	37,790	59,047
28	3,974	5,723	7,790	10,174	12,877	15,897	19,236	22,892	28,262	31,159	34,197	40,697	63,589
30	4,258	6,132	8,346	10,901	13,797	17,033	20,610	24,527	30,280	33,384	36,639	43,604	68,131
32	4,542	6,541	8,902	11,628	14,716	18,168	21,984	26,162	32,299	35,610	39,082	46,511	72,673
34	4,826	6,949	9,459	12,354	15,636	19,304	23,358	27,797	34,318	37,835	41,525	49,418	77,215
36	5,110	7,358	10,015	13,081	16,556	20,439	24,732	29,433	36,336	40,061	43,967	52,325	81,757
38	5,394	7,767	10,572	13,808	17,476	21,575	26,105	31,068	38,355	42,287	46,410	55,231	86,299
40	5,678	8,176	11,128	14,535	18,395	22,710	27,479	32,703	40,374	44,512	48,852	58,138	90,841

Table 6-6. Storage volumes for bulk materials.
Approximate data, based primarily on California Agricultural Circular 517.

Material	lb/ft³	ft³/ton
Alfalfa meal	15	134
Alfalfa, chopped	12	167
Barley meal	28	72
Beet pulp, dried	15	134
Bran	16	125
Brewers grains, dry	15	134
Buckwheat bran	15	134
Buckwheat middlings	23	87
Coconut meal	38	53
Concentrates, typical	45	45
Corn meal	38	53
Ground earcorn, dry	36	56
Cotton seed meal	38	53
Distillers' grain, dry	15	134
Gluten feed	33	61
Gluten meal	46	44
Hominy meal	28	72
Kafir meal	27	74
Lime	60	33
Linseed meal	23	87
Malt sprouts	15	134
Mixed mill feed (bran & middlings)	15	134
Molasses	78	26
Molasses beet pulp	20	100
Meat scraps	34	59
Oats, ground	20	100
Rice bran	20	100
Salt, fine	50	40
Soybean meal	42	48
Tankage	32	63
Wheat bran	14	143
Wheat feed, mixed	15	134
Wheat, ground	45	45
Wheat middlings (Std)	20	100
Wheat screenings	27	74
Pellets, mixed feed	40	50
Pellets, ground hay	45	45

Table 6-7. Capacities of hopper bottom round bulk bins.
Approximate values. Overall height is the distance from ground to roof top and includes 2′ of clearance between ground and hopper bottom. Roof at 35° angle. Bin volume includes roof volume.

Bin dia. ft	Hopper angle degrees	Overall height ft	Total ft³	Total bu	Volume/ft of height* ft³	Volume/ft of height* bu
Sidedraw:						
6	60	15	227	182	28	23
		20	368	295		
		25	510	410		
9	60	22	760	611	64	51
		26	1,015	815		
		30	1,270	1,109		
Centerdraw:						
6	60	10	145	117	28	23
		15	286	230		
		20	428	344		
9	60	18	681	547	64	51
		22	935	751		
		26	1,190	955		
12	60	24	1,613	1,296	113	91
		30	2,292	1,841		
		36	2,970	2,386		
12	45	20	1,493	1,199	113	91
		30	2,624	2,108		
		40	3,755	3,016		
15	45	24	2,739	2,200	177	142
		36	4,860	3,903		
		48	6,980	5,607		
18	45	24	3,512	2,821	255	204
		36	5,268	4,231		
		48	9,619	7,726		
21	45	24	4,191	3,366	346	278
		36	8,348	6,705		
		48	12,504	10,043		

*For calculating volume of different height bins, add or subtract from table value.

Grain Moisture Relationships

Table 6-8. Weight of grain to yield a standard bushel.
A bushel is the amount of grain required to yield: 56 lb of shelled corn, 15.5% moisture content(m.c.); 60 lb of wheat, 13.5% m.c.; 60 lb of soybeans, 13.0% m.c.; 48 lb of barley, 14.5% m.c.; 32 lb of oats, 14.0% m.c.; and 56 lb of rye, 14.0% m.c. Sunflower and grain sorghum are based on 100 lb (cwt) rather than bushels. For sunflower and grain sorghum, table values are the amount of grain to yield 100 lb of sunflower at 10% m.c. and 100 lb of grain sorghum at 14% m.c.
Lb dry matter/bu = 47.32 for corn, 51.90 for wheat, 52.20 for soybeans, 41.04 for barley, 29.24 for oats, and 48.16 for rye. Lb dry matter/cwt = 90 for sunflower and 86 for grain sorghum.

Moisture content, %	Shelled corn	Wheat	Soybeans	Barley	Oats	Rye	Sunflower	Grain sorghum
	lb/bu						lb/cwt base	
5.0	49.81	54.63	54.95	43.20	28.97	50.69	94.73	90.53
5.5	50.07	54.92	55.24	43.43	29.12	50.96	95.24	91.01
6.0	50.34	55.21	55.53	43.66	29.28	51.23	95.74	91.49
6.5	50.61	55.51	55.83	43.89	29.43	51.51	96.26	91.98
7.0	50.88	55.81	56.13	44.13	29.59	51.79	96.78	92.47
7.5	51.16	56.11	56.43	44.37	29.75	52.07	97.30	92.97
8.0	51.44	56.42	56.74	44.61	29.91	52.35	97.83	93.48
8.5	51.72	56.72	57.05	44.85	30.08	52.63	98.36	93.99
9.0	52.00	57.03	57.36	45.10	30.24	52.92	98.90	94.51
9.5	52.29	57.35	57.68	45.35	30.41	53.22	99.45	95.03
10.0	52.58	57.67	58.00	45.60	30.58	53.51	100.00	95.56
10.5	52.87	57.99	58.32	45.85	30.75	53.81	100.56	96.09
11.0	53.17	58.31	58.65	46.11	30.92	54.11	101.12	96.63
11.5	53.47	58.64	58.98	46.37	31.10	54.42	101.69	97.18
12.0	53.77	58.97	59.32	46.64	31.28	54.73	102.28	97.73
12.5	54.08	59.31	59.65	46.90	31.45	55.04	102.86	98.29
13.0	54.39	59.65	60.00	47.17	31.63	55.36	103.46	98.85
13.5	54.71	60.00	60.35	47.45	31.81	55.68	104.05	99.42
14.0	55.02	60.35	60.70	47.72	32.00	56.00	104.66	100.00
14.5	55.35	60.70	61.06	48.00	32.19	56.33	105.26	100.58
15.0	55.67	61.06	61.41	48.28	32.38	56.66	105.89	101.18
15.5	56.00	61.43	61.78	48.57	32.57	57.00	106.52	101.78
16.0	56.33	61.79	62.14	48.86	32.76	57.34	107.15	102.38
16.5	56.67	62.16	62.52	49.15	32.96	57.68	107.79	102.99
17.0	57.01	62.53	62.89	49.45	33.16	58.03	108.44	103.61
17.5	57.36	62.91	63.28	49.75	33.36	58.38	109.10	104.24
18.0	57.71	63.30	63.66	50.05	33.56	58.74	109.76	104.88
18.5	58.06	63.69	64.05	50.36	33.77	59.10	110.44	105.52
19.0	58.42	64.08	64.45	50.67	33.98	59.46	111.11	106.17
19.5	58.78	64.47	64.85	50.98	34.18	59.83	111.81	106.83
20.0	59.15	64.88	65.25	51.30	34.40	60.20	112.51	107.50
21.0	59.90	65.70	66.08	51.95	34.83	60.96	113.93	108.86
22.0	60.67	66.54	66.93	52.62	35.28	61.75	115.39	110.26
23.0	61.45	67.41	67.80	53.30	35.75	62.55	116.89	111.69
24.0	62.26	68.29	68.69	54.01	36.21	63.37	118.43	113.16
25.0	63.09	69.20	69.61	54.73	36.70	64.22	120.00	114.67
26.0	63.95	70.14	70.55	55.47	37.20	65.09	121.62	116.22
27.0	64.82	71.10	71.51	56.22	37.70	65.98	123.29	117.81
28.0	65.72	72.10	72.51	57.00	38.22	66.89	125.00	119.44
29.0	66.65	73.10	73.53	57.80	38.77	67.84	126.77	121.13
30.0	67.60	74.15	74.58	58.63	39.31	68.80	128.58	122.86

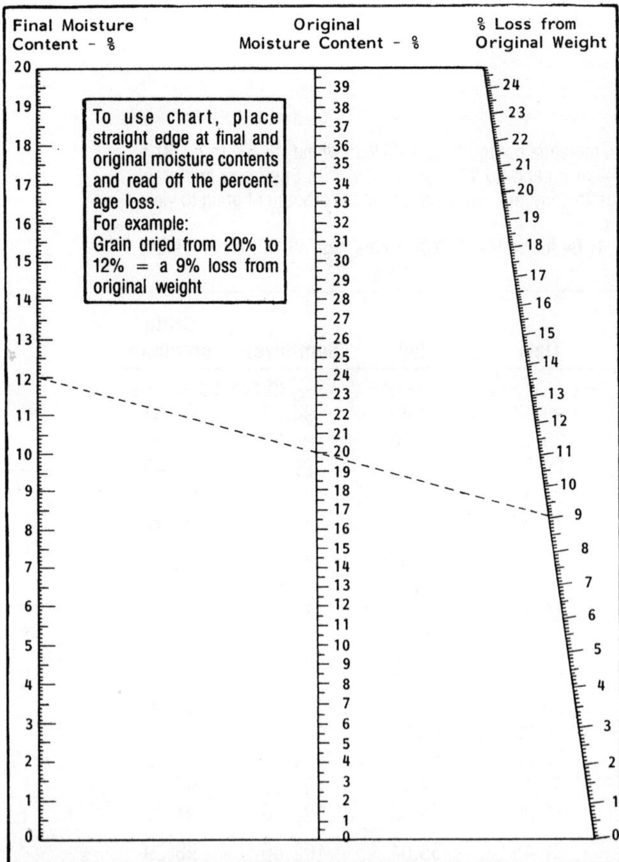

Fig 6-5. Estimating weight loss during drying.
For grain or seed. Chart indicates water loss only. During drying, additional losses (called invisible shrink or handling) are ½% to 1½% of original weight. Add additional losses to chart losses. For example, if the chart loss is 9%, then the total loss is about 9½%-10½%.
Source: University of Georgia College of Agriculture Bulletin N.S. 33.

Table 6-9. Approximate water in ear corn based on kernel moisture.
Values based on No. 2 corn, 15.5% moisture content, 56 lb/bu for shelled corn and 70 lb/bu for ear corn. Kernel dry matter is 47.3 lb/bu, cob dry matter is 11.4 lb/bu, and total ear corn dry matter is 58.7 lb/bu. Moisture contents are determined on a wet basis.
Source: *Principles, Equipment, and Systems for Corn Harvesting*, Agricultural Consulting Associates, Inc., Wooster, Ohio, 1966.

Moisture content Kernel %	Cob %	Water in Kernels lb/bu	Cobs lb/bu	Ear lb/bu
10	8.6	5.3	1.1	6.3
12	11.7	6.5	1.5	8.0
14	15.5	7.7	2.1	9.8
15.5	18.6	8.7	2.6	11.3
16	20.0	9.0	2.9	11.9
18	26.2	10.4	4.1	14.4
20	33.8	11.8	5.8	17.7
22	40.0	13.4	7.6	21.0
24	44.3	14.9	9.1	24.0
26	48.0	16.6	10.5	27.2
28	50.6	18.4	11.7	30.1
30	52.8	20.3	12.8	33.0
35	56.4	25.5	14.7	40.2

Example 6-4:

Determine the weight loss as a percentage of original weight when grain is dried from 20% to 12% moisture content.

Solution:

Using Fig 6-5, weight loss is 9%.

Example 6-5:

Determine the amount of water removed when drying shelled corn from 26% to 15½% moisture content.

Solution:

From Table 6-8, 63.95 lb of 26% m.c. shell corn is needed to yield a bushel of 15½% m.c. shelled corn (56 lb). The amount of water removed is:

63.95 lb − 56.0 lb = 7.95 lb of water

Harvesting Capacity

To estimate harvesting capacity use Eq 6-1:

Eq 6-1.

$$HC = (S \times W \times EFF \times Y) \div 8.25$$

HC = harvesting capacity, bu/hr
S = combine speed, mph
W = combine harvesting width, ft
EFF = harvesting efficiency = 0.65 to 0.85
Y = yield, bu/acre

Harvest efficiency considers downtime for unloading combines, making adjustments, and turning. Use a harvest efficiency of:

- 0.85 for long fields and no stopping to unload
- 0.75 for long fields and stopping or short fields requiring more turning and no stopping
- 0.65 for short fields and stopping to unload.

Example 6-6:

Determine the harvesting capacity for combining corn. The combine travels 4 mph, has a head for four 38″ rows, and is stopped to unload. Yield is 150 bu/acre.

Solution:

1. Assume a harvesting efficiency of 0.75, because harvesting is stopped to unload.
2. Calculate the combine harvesting width.

   W = 4 rows × 38″/row ÷ 12″/ft = 12.67′

3. Using Eq 6-1, the harvesting capacity is:

   HC = (4 mph × 12.67 ft × 0.75 × 150 bu/A) ÷ 8.25 = 691 bu/hr

Airflow Static Pressure

Table 6-10. Static pressures for airflow through grain.
Derived from ASAE Standard D245.4,1985. For grain only, does not include pressure losses due to duct entrances, transitions, or grain-perforated floor interface. Neglect losses for full drying floors with open area greater than 7% and air velocities less than 2,000 fpm in the transition. Use table values directly. For duct or pad type aeration systems and duct velocity less than 2,000 fpm, add ½" of static pressure to the table value. If duct or transition velocity is greater than 2,000 fpm, contact an engineer to determine the static pressure. A multiplier factor is used to convert the pressure drop of clean unpacked grain as given by ASAE Data: ASAE D272 to field conditions.

Grain type	Grain depth ft	Airflow rate cfm/bu ¹⁄₂₀	¹⁄₁₀	⅕	½	¾	1	1¼	1½
		- - - - - - - - - - - in. of water column - - - - - - - - - - -							
Barley	10			0.3	0.7	1.1	1.3	2.0	2.6
(Multiplier factor = 1.5)	15		0.3	0.6	1.6	2.7	3.8	5.0	6.4
Oats*	20		0.5	1.1	3.0	5.0	7.3	9.8	12.4
Sunflower*	25	0.3	0.8	1.7	5.1	8.4	12.0	16.3	
	30	0.6	1.2	2.5	7.6	12.7			
	35	0.8	1.6	3.5	10.8				
	40	1.0	2.2	4.7	14.6				
	50	1.6	3.5	7.7					
	60	2.4	5.1	11.5					
	70	3.2	7.0						
	80	4.3	9.5						
	90	5.5	12.2						
	100	6.9							
Shelled corn	10			0.1	0.3	0.5	0.8	1.1	1.4
(Multiplier factor = 1.5)	15		0.1	0.3	0.8	1.4	2.0	2.8	3.6
	20		0.2	0.5	1.6	2.7	4.1	5.6	7.4
	25		0.4	0.8	2.6	4.7	7.1	9.9	12.9
	30		0.5	1.2	4.1	7.2	11.2	15.5	
	35	0.4	0.8	1.7	5.9	10.6			
	40	0.5	1.0	2.3	8.1				
	50	0.7	1.6	3.9	14.1				
	60	1.1	2.5	5.9					
	70	1.5	3.5	8.6					
	80	2.0	4.7	11.9					
	90	2.6	6.1						
	100	3.3	7.8						
Soybeans	10				0.2	0.4	0.5	0.7	0.8
(Multiplier factor = 1.3)	15			0.2	0.6	0.9	1.3	1.7	2.2
	20		0.2	0.4	1.0	1.7	2.4	3.6	4.6
	25		0.3	0.6	1.7	2.8	4.2	5.7	7.5
	30	0.2	0.4	0.9	2.5	4.3	6.5	9.0	11.7
	35	0.3	0.6	1.2	3.6	6.2	9.3	13.1	
	40	0.4	0.8	1.6	4.9	8.6	13.0		
	50	0.6	1.2	2.6	9.3				
	60	0.8	1.7	3.9	13.7				
	70	1.1	2.4	5.4					
	80	1.5	3.2	7.3					
	90	1.9	4.1	9.5					
	100	2.4	5.2	12.2					
Wheat	10			0.4	1.1	1.6	2.3	3.0	3.7
(Multiplier factor = 1.3)	15		0.5	0.9	2.5	4.0	5.6	7.2	9.2
Grain sorghum*	20		0.8	1.7	4.6	7.4	10.4		
	25	0.7	1.3	2.7	7.5	12.0			
	30	1.0	1.4	3.9	11.1				
	35	1.3	2.6	5.4					
	40	1.7	3.3	7.1					
	50	2.6	5.3	11.4					
	60	3.7	7.7						
	70	5.1	10.6						
	80	6.7	14.1						
	90	8.5							
	100	10.7							

*Limited research results available for oats, sunflower, and grain sorghum. Values for barley can be used as a good estimate for oats and sunflower. Wheat values can be used for grain sorghum.

Sizing Motors, Pulleys, and Belts

Motors

Calculate motor hp required for equipment by multiplying the hp/10′ values from the appropriate tables in this book (2-3, 2-7, 2-8) by the conveyor length/10′ value. Next choose the motor hp from Table 6-11. See Example 2-2.

Electric motors mounted with the shaft in positions other than horizontal must have thrust bearings. Operate ½ hp and larger electric motors on 240 volt service (nominal) and control them with a magnetic motor-starting switch. Attach the equipment grounding wire to the electric motor frame for shock protection. Make sure overcurrent protection is properly sized for the motor and circuit. If an on-off switch is used for small motors, install time delay fusetrons or fusestats sized to 115% of the nameplate operating current.

Table 6-11. Motor selection for continuous conveyor operation.

Calculated conveyor hp	Electric motor hp	Gasoline engine hp
Up to 0.27	¼	½
0.28 to 0.35	⅓	⅔
0.36 to 0.55	½	1
0.56 to 0.81	¾	1½
0.82 to 1.10	1	2
1.11 to 1.60	1½	3
1.61 to 2.10	2	4
2.11 to 3.20	3	5
3.21 to 5.25	5	8
5.26 to 7.90	7½	12
7.91 to 10.50	10	15

Pulleys

Most electric drive motors used to power conveyors run at 1,725 rpm. Gas engines run at relatively higher rpm. Usually, equipment needs to run slower than the motor. Use different size pulleys on motors and equipment to adjust speed. The motor pulley is usually smaller than the equipment pulley. For electric motors ½ hp or larger, use at least a 3″ outside diameter motor pulley.

To determine pulley sizes, choose a pulley size for the motor, then calculate the equipment pulley size. Pulley size calculations are based on pitch diameter, not outside diameter, Fig 6-6. Use Eq 6-2 to calculate equipment pulley size.

Eq 6-2.

$$EP = (MRPM \times MP) \div ERPM$$

EP = equipment pulley pitch diameter, in.
MRPM = motor rpm, revolutions per minute, usually 1,725 rpm but check the motor nameplate.
MP = motor pulley pitch diameter, in.
ERPM = desired equipment rpm.

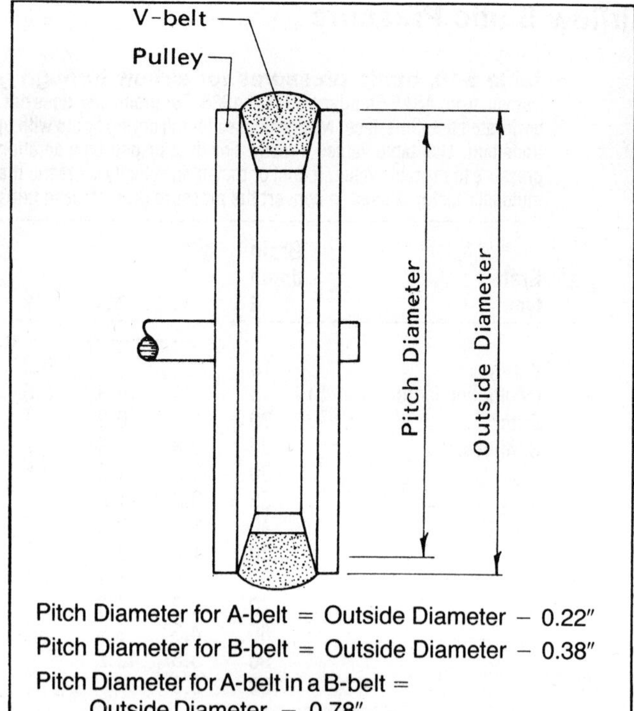

Pitch Diameter for A-belt = Outside Diameter − 0.22″
Pitch Diameter for B-belt = Outside Diameter − 0.38″
Pitch Diameter for A-belt in a B-belt = Outside Diameter − 0.78″

Fig 6-6. Outside and pitch diameter of a pulley.

Belts

V-belts are classified according to cross-sectional size, Fig 6-7. Belt type and the number of belts required for a piece of equipment depends on the horsepower being transmitted, Table 6-12. If belts are too small or not enough belts are used, they will slip and wear out quickly. Select pulleys with the same width as the belt or belts, Fig 6-7. This is particularly important if equipment is to operate at the calculated speed.

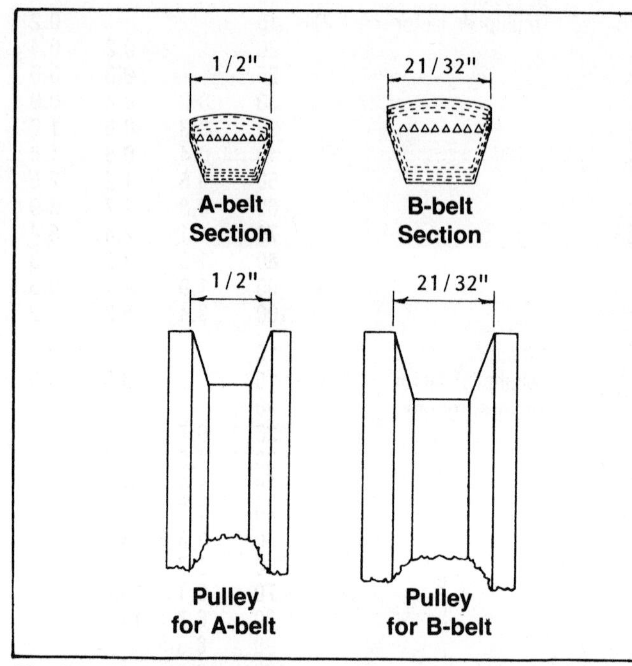

Fig 6-7. Typical V-belts and pulleys for grain equipment.

Table 6-12. Required number and type of V-belts.
When more than 3 belts are required, consider a chain drive.

Motor pulley outside dia., in.	Motor hp								
	½ or under	¾	1	1½	2	3	5	7½	10
	- - - number and type of belt(s) required - - -								
3	1-A	1-A	1-A	2-A	2-A	3-A	5-A	-	-
3½	1-A	1-A	1-A	2-A	2-A	3-A	4-A	-	-
4	1-A	1-A	1-A	1-A	2-A	2-A	3-A	5-A	-
4½	1-A	1-A	1-A	1-A	1-A	2-A	3-A	5-A	5-A
5	1-A	1-A	1-A	1-A	1-A	2-A	3-A	4-A	5-A
5½	1-A	1-A	1-A	1-A	1-A	1-B	2-B	3-B	4-B
6	1-A	1-A	1-A	1-A	1-A	1-B	2-B	2-B	3-B
7	1-A	1-A	1-A	1-A	1-A	1-B	2-B	2-B	3-B
8	1-A	1-A	1-A	1-A	1-A	1-B	1-B	2-B	2-B

Example 6-7:

A 6″ auger has a 5 hp motor with a 4″ pulley that operates at 1,725 rpm. Determine the belts and auger pulley size required to operate at 600 rpm.

Solution:

1. From Table 6-12, three A belts are required to transmit 5 hp with a 4″ motor pulley.
2. Calculate the motor pulley pitch diameter. From Fig 6-6, subtract 0.22 from the outside diameter.

   Pitch dia. = 4″ − 0.22″ = 3.78″
3. Calculate the auger pulley pitch diameter using Eq 6-2.

   EP = (1,725 rpm × 3.78″) ÷ 600 rpm = 10.87″
4. The outside diameter of the auger pulley is:

   10.87″ + 0.22″ = 11.09″

   Select an 11″ auger pulley.

Conversion Factors for Grain Systems

Table 6-13. Bulk grain handling units.

Physical quantity	English Unit	English Symbol	Metric Unit	Metric Symbol
Length	foot	ft	meter	m
Weight	pound	lb	kilogram	kg
Land area	acres	acre	hectare	ha
Volume	cubic feet	ft³	cubic meter	m³
Flowrate	cubic feet per min.	ft³/min	cubic meter per second	m³/s
Static pressure	inches of water	in.	kilopascal	Kpa
Test weight	pound per bushel	lb/bu	kilograms per hectoliter	kg/hl
Crop yield	bushel per acre	bu/acre	kilograms per hectare	kg/ha
Production	bushel	bu	metric ton	t
Unit of trade	bushel	bu	metric ton	t

Table 6-14. Conversions.
Multiply to the right: acres × 43,560 = ft².
Divide to the left: ft² ÷ 43,560 = acres.

Unit	Times	Equals
Acres	43,560	ft²
	4,840	yd²
	160	square rods
	1/640	square mile
Bushels	1.245	ft³
	2.5	ft³ ear corn
ft³	1728	in³
	0.4	bu ear corn
	0.8	bu grain
Cubic yard	27	ft³
concrete	81	ft² of 4″ floor
concrete	54	ft² of 6″ floor
Miles	5,280	ft
	1,760	yd
	320	rods
Pressure, psi	2.31	ft of water head
	27.69	in of water head
Rods	16.5	ft
	5.5	yd

Table 6-15. Metric conversions.
Multiply to the right: in. × 2.54 = cm
Divide to the left: cm ÷ 2.54 = inches

Unit	Times	Equals
Length		
inches	2.540	cm
feet	0.3048	m
yards	0.9144	m
miles	1.609	km
Area		
in²	6.451	cm²
ft²	0.09290	m²
yd²	0.8361	m²
acres	0.4047	hectares
acres	4047.0	m²
mile²	2.590	km²
Volume		
in³	16.39	cm³
in³	0.01639	liters (L)
ft³	0.02832	m³
cubic yards	0.7646	m³
bushel	0.03524	m³
Mass		
pounds	0.4536	kg
ton (2,000 lb)	907.2	kg
ton (2,000 lb)	0.9072	tonne (t)
Velocity		
ft/sec	0.3048	m/s
miles/hour (mph)	1.609	km/h
Flowrate		
cfm	0.0004719	m³/s
Pressure		
in. of water	0.2488	KPa
Temperature		
Fahrenheit (F); Celsius (C)		
C = (F − 32) ÷ 1.8		
F = (C × 1.8) + 32		

Table 6-16. Grain production conversions. A standard bushel is 1.245 ft^3 by volume. Because different grains have different standard bushel definitions by weight, a special conversion factor is used for each crop. Values based on lb/bu from Table 6-4 and 2,200 lb/metric ton.

Grain	Bushels to metric tons	Metric tons to bushels
	Multiply by	
Corn	0.0255	39.286
Soybeans	0.0273	36.667
Wheat	0.0273	36.667
Oats	0.0145	68.750
Rapeseed	0.0227	44.000
Barley	0.0218	45.833
Rye	0.0255	39.286
Flaxseed	0.0255	39.286

Table 6-17. Yield conversions.

Grain	bu/ac to kg/ha	kg/ha to bu/ac
	Multiply by	
Barley	53.81	0.0186
Corn	62.78	0.0159
Flaxseed	62.78	0.0159
Oats	35.87	0.0279
Rapeseed	56.05	0.0178
Rye	62.78	0.0159
Soybeans	67.26	0.0149
Wheat	67.26	0.0149

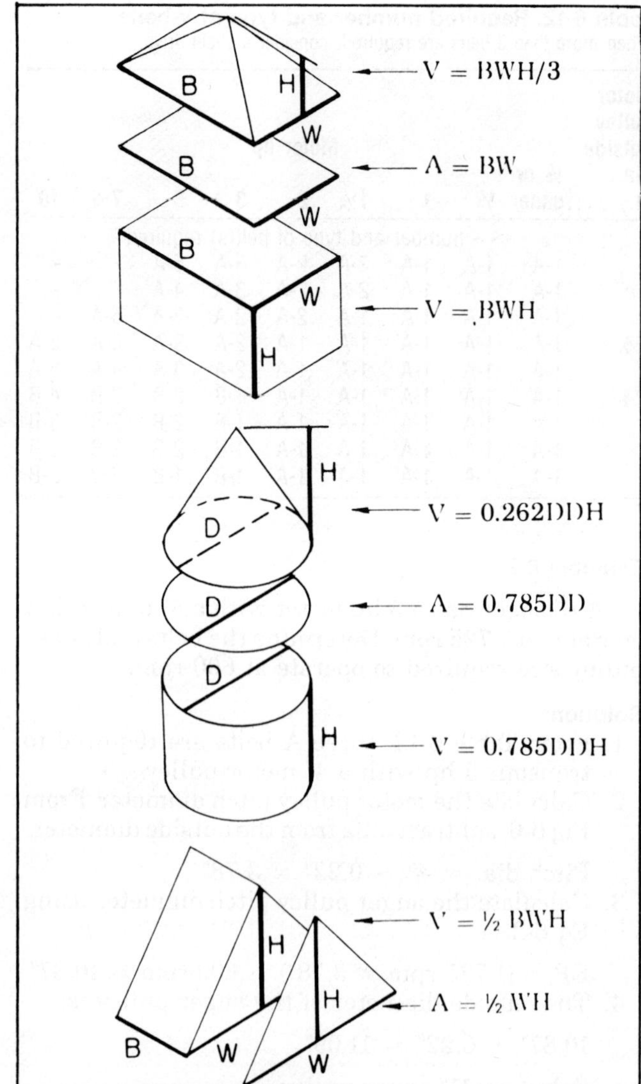

Fig 6-8. Areas and volumes.

7. GLOSSARY

Aeration: Process of moving air through stored grain to control grain temperature.

Ambient air: Outdoor or surrounding air such as the air around a bin.

AMCA: Air Movement and Control Association; an independent organization responsible for fan testing, specifications, and procedures.

Automatic/Automation: A process (i.e. drying or conveying) controlled with electrical switches.

BCFM: Broken corn and foreign material for shelled corn. Generally referred to as fines in this book.

Biomass: Plant or organic material used as a fuel.

Cash grain: Grain that is grown and processed for sale rather than feeding.

Cfm: Cubic feet per minute (ft^3/min); unit of airflow.

Cfm per bu: Cubic feet per minute per bushel; unit of airflow per unit of volume.

Cooling bin: Separate bin used to cool grain dried with a high temperature dryer.

Dryeration: The process of cooling grain from a high temperature dryer after the grain has steeped for at least 4 hr.

Duct: Circular or rectangular tube that air flows or moves through in a grain system.

Fines/Dockage: Broken kernels, foreign material, cobs, weed seeds, etc. intermixed in the grain. Fines refers to corn; dockage to small grains.

Fpm: Feet per minute (ft/min); unit of air velocity.

ft^2: Square feet; unit of area.

ft^3: Cubic feet; unit of volume.

Grain center: The grain drying, handling, and storage system. Includes equipment or structures for grain receiving, conveying, drying, cooling, storage, and loadout.

Grain conditioning: Drying grain or cooling hot grain.

Gravity flow: Non-mechanical flow; results from a difference in elevation between the discharge of a holding bin and receiving or the inlet and outlet of a bucket elevator transfer tube.

High temperature (high speed) dryer: A dryer that dries grain relatively fast with air heated to at least 120 F.

Holding bin: A bin used to hold grain before it is dried or loaded out.

Hub: The grain center receiving and/or loadout area, often includes a bucket elevator.

Low temperature dryer: A dryer that slowly dries grain with unheated air or air heated up to 10 F.

LP gas: Liquified petroleum gas.

Management center building: A building that houses controls, testing equipment, and records. It can include an office, grain receiving/loadout, and feed processing.

Manometer: Instrument for measuring static pressure.

Metric tonne: Metric unit of weight; 1 metric tonne = 2,200 lb.

Moisture content: The percent of water by weight. In this book, it is measured on a wet basis.

Pack factor: A factor used to take into account the increase in static pressure or grain resistance to air flow commonly found in bins. This is not to be confused with the packing factor used to determine storage capacity of bins.

Pit: Below-ground grain receiving container.

Refuse: Plant or biomass material used for fuel.

Rpm: Revolutions per minute—usually refers to the rotating speed of shafts, pulleys, and augers.

Screenings: Fines and other material separated from grain in the screening process.

Self-contained dryer: A unit that only dries and cools grain. Has little or no storage capacity.

Static pressure: Resistance that needs to be overcome to move air through a grain system. Usually measured in inches of water column.

Steep: To hold grain taken directly from a high temperature dryer without cooling in a bin for 4 to 12 hr to permit the moisture and temperature to equalize within the grain. Used with in-storage cooling and dryeration to increase grain quality, increase drying capacity, and lower drying costs.

Stress cracks: Fractures inside the kernel that may not come through the outer seed coat.

Surge bin: A small bin to hold grain before being conveyed.

Test weight: The weight of a standard bushel of grain (1.245 ft^3) used as an official measure for grading purposes.

Transition: Usually a tapered duct between a fan and distribution duct or drying bin.

8. RELATED REFERENCES

Available from the Extension Agricultural Engineer at any of the institutions listed on the inside front cover or from Midwest Plan Service.

Handbooks and Digests

MWPS-1 *Structures and Environment Handbook.*

MWPS-2 *Farmstead Planning Handbook.*

MWPS-22 *Low Temperature & Solar Grain Drying Handbook.*

MWPS-28 *Farm Building Wiring.*

AED-7 *Grain Bin Floors & Foundations.*

AED-12 *Remodeling Corn Cribs for Small Grain Storage.*

AED-20 *Managing Dry Grain in Storage.*

AED-26 *Farm and Home Concrete*

AED-28 *Cast-in-Place Concrete Walls for Farm and Home*

AED-29 *Aeration Systems for Grain Storage* (available May 1988)

Plans

mwps-73210 *Movable Grain Storage Walls.*

mwps-73217 *Small Feed Bins.*

mwps-73220 *48′ Pole Grain Storage.*

mwps-73250 *Small Storage Buildings.*

mwps-73258 *10,000 Bushel Gable Grain Storage.*

mwps-73260 *15,000 Bushel Gable Grain Storage.*

mwps-73292 *Grain-Feed Handling Center, Work Tower Across Drive, Equal Sheds.*

mwps-73293 *Grain-Feed Handling Center, Work Tower Across Drive, One-Way Shed.*

mwps-73294 *Grain-Feed Handling Center, Work Tower Beside Drive, Offset Gable.*

9. INDEX